RESIDUE REVIEWS

VOLUME 65

Effects of triazine herbicides on
the physiology of plants

SINGLE PESTICIDE VOLUME:

EFFECTS OF TRIAZINE HERBICIDES ON THE PHYSIOLOGY OF PLANTS

RESIDUE REVIEWS

Residues of Pesticides and Other Contaminants in the Total Environment

Editor
FRANCIS A. GUNTHER

Assistant Editor
JANE DAVIES GUNTHER

Riverside, California

VOLUME 65

Springer Science+Business Media, LLC
1976

Originally published by Springer-Verlag New York, Inc. in 1976
Softcover reprint of the hardcover 1st edition 1976

Library of Congress Catalog Card Number 62-18595.

ISBN 978-3-662-37686-7　ISBN 978-3-662-38488-6 (eBook)
DOI 10.1007/978-3-662-38488-6

Foreword

Worldwide concern in scientific, industrial, and governmental communities over traces of toxic chemicals in foodstuffs and in both abiotic and biotic environments has justified the present triumvirate of specialized publications in this field: comprehensive reviews, rapidly published progress reports, and archival documentations. These three publications are integrated and scheduled to provide in international communication the coherency essential for nonduplicative and current progress in a field as dynamic and complex as environmental contamination and toxicology. Until now there has been no journal or other publication series reserved exclusively for the diversified literature on "toxic" chemicals in our foods, our feeds, our geographical surroundings, our domestic animals, our wildlife, and ourselves. Around the world immense efforts and many talents have been mobilized to technical and other evaluations of natures, locales, magnitudes, fates, and toxicology of the persisting residues of these chemicals loosed upon the world. Among the sequelae of this broad new emphasis has been an inescapable need for an articulated set of authoritative publications where one could expect to find the latest important world literature produced by this emerging area of science together with documentation of pertinent ancillary legislation.

The research director and the legislative or administrative advisor do not have the time even to scan the large number of technical publications that might contain articles important to current responsibility; these individuals need the background provided by detailed reviews plus an assured awareness of newly developing information, all with minimum time for literature searching. Similarly, the scientist assigned or attracted to a new problem has the requirements of gleaning all literature pertinent to his task, publishing quickly new developments or important new experimental details to inform others of findings that might alter their own efforts, and eventually publishing all his supporting data and conclusions for archival purposes.

The end result of this concern over these chores and responsibilities and with uniform, encompassing, and timely publication outlets in the field of environmental contamination and toxicology is the Springer-Verlag (Heidelberg and New York) triumvirate:

Residue Reviews (vol. 1 in 1962) for basically detailed review articles concerned with any aspects of residues of pesticides and other chemical contaminants in the total environment, including toxicological considerations and consequences.

Bulletin of Environmental Contamination and Toxicology (vol. 1 in 1966) for rapid publication of short reports of significant advances and discoveries in the fields of air, soil, water, and food contamination and pollution as well as methodology and other disciplines concerned with the introduction, presence, and effects of toxicants in the total environment.

Archives of Environmental Contamination and Toxicology (vol. 1 in 1973) for important complete articles emphasizing and describing original experimental or theoretical research work pertaining to the scientific aspects of chemical contaminants in the environment.

Manuscripts for *Residue Reviews* and the *Archives* are in identical formats and are subject to review, by workers in the field, for adequacy and value; manuscripts for the *Bulletin* are not reviewed and are published by photo-offset to provide the latest results without delay. The individual editors of these three publications comprise the Joint Coordinating Board of Editors with referral within the Board of manuscripts submitted to one publication but deemed by major emphasis or length more suitable for one of the others.

September 1, 1976 Coordinating Board of Editors

Preface

That residues of pesticide and other contaminants in the total environment are of concern to everyone everywhere is attested by the reception accorded previous volumes of "Residue Reviews" and by the gratifying enthusiasm, sincerity, and efforts shown by all the individuals from whom manuscripts have been solicited. Despite much propaganda to the contrary, there can never be any serious question that pest-control chemicals and food-additive chemicals are essential to adequate food production, manufacture, marketing, and storage, yet without continuing surveillance and intelligent control some of those that persist in our foodstuffs could at times conceivably endanger the public health. Ensuring safety-in-use of these many chemicals is a dynamic challenge, for established ones are continually being displaced by newly developed ones more acceptable to food technologists, pharmacologists, toxicologists, and changing pest-control requirements in progressive food-producing economies.

These matters are of genuine concern to increasing numbers of governmental agencies and legislative bodies around the world, for some of these chemicals have resulted in a few mishaps from improper use. Adequate safety-in-use evaluations of any of these chemicals persisting into our foodstuffs are not simple matters, and they incorporate the considered judgments of many individuals highly trained in a variety of complex biological, chemical, food technological, medical, pharmacological, and toxicological disciplines.

It is hoped that "Residue Reviews" will continue to serve as an integrating factor both in focusing attention upon those many residue matters requiring further attention and in collating for variously trained readers present knowledge in specific important areas of residue and related endeavors involved with other chemical contaminants in the total environment. The contents of this and previous volumes of "Residue Reviews" illustrate these objectives. Since manuscripts are published in the order in which they are received in final form, it may seem that some important aspects of residue analytical chemistry, biochemistry, human and animal medicine, legislation, pharmacology, physiology, regulation, and toxicology are being neglected; to the contrary, these apparent omissions are recognized, and some pertinent manuscripts are in preparation. However, the field is so large and the interests in it are so varied that the editors and the Advisory Board earnestly solicit suggestions of topics and authors to help make this international book-series even more useful and informative.

"Residue Reviews" attempts to provide concise, critical reviews of timely advances, philosophy, and significant areas of accomplished or needed endeavor in the total field of residues of these and other foreign chemicals in any segment of the environment. These reviews are either general or specific, but properly they may lie in the domains of analytical chemistry and its methodology, biochemistry, human and animal medicine, legislation, pharmacology, physiology, regulation, and toxicology; certain affairs in the realm of food technology concerned specifically with pesticide and other food-additive problems are also appropriate subject matter. The justification for the preparation of any review for this book-series is that it deals with some aspect of the many real problems arising from the presence of any "foreign" chemicals in our surroundings. Thus, manuscripts may encompass those matters, in any country, which are involved in allowing pesticide and other plant-protecting chemicals to be used safely in producing, storing, and shipping crops. Added plant or animal pest-control chemicals or their metabolites that may persist into meat and other edible animal products (milk and milk products, eggs, etc.) are also residues and are within this scope. The so-called food additives (substances deliberately added to foods for flavor, odor, appearance, etc., as well as those inadvertently added during manufacture, packaging, distribution, storage, etc.) are also considered suitable review material. In addition, contaminant chemicals added in any manner to air, water, soil or plant or animal life are within this purview and these objectives.

Manuscripts are normally contributed by invitation but suggested topics are welcome. Preliminary communication with the editors is necessary before volunteered reviews are submitted in manuscript form.

Department of Entomology
University of California
Riverside, California
September 1, 1976

F.A.G.
J.D.G.

Effects of triazine herbicides on the physiology of plants

By

E. EBERT* and S. W. DUMFORD**

Contents

---

* Agrochemicals Division, Ciba-Geigy Ltd., Basle, Switzerland.
** Agricultural Division, Ciba-Geigy Corp., Charlotte, N. C.
*** Procida, Marseille, France.
**** Armentieres, 106, rue Jules Ferry, France.

Foreword

Soon after the discovery of the potent herbicidal activity of certain classes of *N*-alkylated diamino-*s*-triazines, it was recognized that they act through interference with photosynthesis. Photosynthesis has been and is still the center of the interest of plant physiologists, biochemists, and agronomists. It is not surprising, therefore, that *s*-triazines met quickly the attention of a large number of scientists engaged in this argument. Broad practical use of the compounds, moreover, stimulated work on their influence on plant respiration and transpiration, on the plant constituents, on plant morphology and growth, on the seed, and on the reasons for plant variability in resistance.

Some of the aspects mentioned have been reviewed already; however, no effort has been made up to now to gather together the results from the entity of the worldwide research effort documented in a great number of reports, publications, and theses. The authors of this volume merit, therefore, to be acknowledged for filling this gap. Their reviewing and their comments benefit from the fact that their active involvement in basic studies had kept them in touch with the respective scientific community for a long time. The dimension of their undertaking is illustrated by more than 700 references quoted herein.

The picture of the acquired knowledge shows in an impressive way the high degree of cross-fertilization between fundamental studies and the close observation of the phenomena occuring in the field.

Although much is known on the triazine activity now, more is still open for exploration, so especially the mode of action on the molecular level. This review will be welcome to all those who are and will be challenged by the fascinating objective of better understanding. Future research based on the knowlege of the past experience will allow them to reach new horizons.

Enrico Knuesli

Ciba-Geigy Ltd.
Basle, Switzerland May, 1975

I. Photosynthesis

Many chemicals, some having herbicidal properties, have been shown to interfere with at least one of the steps of photosynthesis. Alterations of this vital and unique process in green organisms result in injury or death of plants. Among the different reactions involved in the photosynthetic process, the Hill reaction is inhibited by many herbicides. The Hill reaction is defined as photosynthetic oxygen evolution and reduction of an electron acceptor by the photosynthetic transport chain, *i.e.*, as the light-induced electron transfer in chloroplasts from water (resulting in its oxidation to oxygen) to an added, artificial or physiological, electron acceptor (TREBST 1972). This reaction is inhibited by phenylureas, *N*-phenylcarbamates, *s*-triazines, acylanilides, uracils, some halogenated benzonitriles, imidazoles and polycyclic ureas, trifluoromethyl-benzimidazoles, pyridazinones, and aminotriazinones. Other herbicides or potential herbicides, like benziminazoles, carbonylcyanidphenyl-hydrazones, and anilinothiophenes are inhibitors or uncouplers of the photophosphorylation coupled to electron transport in the Hill reaction. The action of these or other inhibitors and the necessary general background of photosynthesis was reviewed by GOOD (1961), GOOD and IZAWA (1964), GOOD *et al.* (1966), AVRON (1966), BARTH and MICHEL (1969), BUECHEL (1972), DRABER *et al.* (1974), and TREBST (1970 and 1974), with special attention paid to the Hill reaction (MORELAND 1967 and 1969).

a) Triazine action in photosynthesis

From the early work of MORELAND and co-workers (MORELAND and HILL 1962, MORELAND *et al.* 1959, EXER 1958 and 1961), it is known that the Hill reaction with ferricyanide, Janus green, or other artificial electron acceptors by isolated chloroplasts of barley or spinach was drastically reduced by simazine. Many researchers have shown this inhibition with several triazines (EXER 1958 and 1961, GOOD 1961, GYSIN and KNUESLI 1960 a and b, MORELAND *et al.* 1959, MORELAND and HILL 1962) in isolated chloroplasts or in algae (BRAGINSKI *et al.* 1963, HOLLISTER and WALSH 1973, ZWEIG *et al.* 1963, WALSH 1972), as well as with the other herbicides mentioned above (see reviews there). In studies with intact plants, uptake of CO_2 was reduced by simazine, atrazine, simetone, ametryn, atraton, and metoprotryn in a wide series of terrestrial (ASHTON *et al.* 1960; BOLHAR-NORDENKAMPF 1970; COUCH 1966; COUCH and DAVIS 1966; FUNDERBURK and CARTER 1965; GYSIN and KNUESLI 1960; HARTISCH 1967; IMBAMBA 1970; IMBAMBA and MOSS 1971; MASHTAKOV and PROHORCIK 1966; OLECH 1968; VAN OORSCHOT 1964, 1965, 1968 and 1970 a and b; ROTH 1959; SASAKI and KOZLOWSKI 1965 and 1967 b; SIKKA and DAVIS 1969; SMITH and ILNICKI 1972 a and b; SPESIVTSEV *et al.* 1969 a and b; TIESZEN 1970; ZWEIG and ASHTON 1962) and aquatic plants (RAU and GRIMME 1971, SUTTON *et al.* 1969) which were crops and weeds either tolerant or susceptible to triazine herbicides. This inhibition of CO_2-uptake was followed by a sharp decrease of the total sugar content in the plant (KUZNETSOV 1966, LIEPINA 1969, MASHTAKOV and PROKHORCIK 1962 a and 1966, SWIETOCHOWSKI *et al.* 1966, TIMOFEVA 1967). Starch disappeared from the leaves of *Coleus* after triazine treatment as demonstrated by GAST (1958). Phytotoxicity symptoms from triazine application resulting from depletion of carbohydrates were overcome by glucose in simazine-treated barley plants (MORELAND *et al.* 1959). Also sucrose suppressed the disappearance of starch in *Coleus* leaves treated with triazines (GAST 1958). In wheat seedlings, the effects of atrazine, simazine, and prometryn on the enzyme activity of ribulose diphosphate carboxylase were investigated, which is considered to be a key enzyme during CO_2-assimilation. This enzyme was not influenced by the triazines (DECLEIRE and BASTIN 1973).

In a recent study, VAN OORSCHOT (1974) investigated the effect of CO_2-free air and CO_2-free nitrogen on the initiation of phytotoxicity symptoms in *Phaseolus vulgaris*. He found that CO_2-free air increased the degree of photosynthetic inhibition in the plants with exposure time and with light intensity. CO_2-free nitrogen delayed the beginning of injury symptoms. From his results he concluded that a lack of carbohydrates is not the only cause of the herbicide action.

These results together with the inhibition of the Hill reaction by *s*-triazines suggested that the action of the herbicide is located at the early steps in the photochemical conversion of energy during the photosynthetic process but before the biosynthesis of saccharides. The exact location of the inhibitory site of most of the herbicides has developed from studies with the herbicide DCMU which have been extended to triazines, namely:

—photoreduction and CO_2-reduction of H_2-adapted algae were not inhibited by amino-triazines (BISHOP 1962),

—O_2-evolution, photoreduction and photophosphorylation were inhibited by *s*-triazines with ferricyanide or FMN as electron acceptors (GOOD 1961),

—PMS-mediated cyclic phosphorylation was not sensitive to triazines (GOOD 1961), and

—noncyclic photophosphorylation can be restored by an electron donor for photosystem I.

Therefore, inhibition of CO_2-evolution, noncyclic electron flow, and phosphorylation occur only when water is the electron donor. Consequently, the mechanism of inhibition involved should be the one responsible for the oxidation of water to molecular oxygen (BISHOP 1962, GOOD 1961, HILTON *et al.* 1963). Reviews and detailed discussion of primary processes and energy transfer have been presented by FORK and AMESZ (1969), HOCH and KNOX (1968), KOK and CHENIAE (1966), and more recently by CHENIAE (1970) and TREBST (1974).

The physiological and biochemical studies on the location of the inhibitory site of the herbicides in the photosynthetic water-splitting reaction suggested an inhibition of photosynthetic electron transport close to photosystem II.

Biophysical studies on fluorescence emission, action spectra, intensity yield, life time and polarization, and fast oxygen evolution measurements gave detailed information on the reaction mechanism and sequence of carriers around photosystem II including information on energy transfer, distribution of pigments within the photochemical centers, and action of added chemicals. In the presence of an artificial electron acceptor, fluoresence of algae at a wavelength above 650 nm under incident light ($\lambda = 475$ nm) presents the following pattern: first, fluorescence rises rapidly after the light is turned on, then increases to a maximum, and finally drops slowly to reach a steady state (LAVOREL 1959 and 1962). Simazine and atrazine were found to increase the fluorescence intensity of *Chlorella*, and no decline followed the rise of fluorescence (ZWEIG *et al.* 1963). The magnitude of increase of fluorescence intensity paralleled the inhibitor potencies of the 2 chemicals on oxygen evolution, *i.e.*, the pI_{50} correspond to the pF_{50} (quantitative evaluation of 50% fluorescence increase). In spinach chloroplasts, MURATA *et al.* (1966 a and b) reported a slightly different situation: no maximum of fluorescence appeared in the transient phase, and fluorescence intensity reached a steady state without any previous rise; simazine had different effects on the time course of fluorescence, *i.e.*, prolongation of the transient phase at low concentrations and its shortening at high concentrations. However, no interpretation was given of this last point.

Since atrazine and simazine change the time-course of fluorescence, this would mean either a direct interaction with chlorophyll in the reaction center II, presumably at the acceptor site of photosystem II, or some interference with the donor site. In the first case triazines would interfere with the flow of electrons from water to excited chlorophyll molecules which will then remain with their electron holes unfilled and will bleach (VAN OVERBEEK 1962). This would also explain the increase of toxicity which parallels the increase of light intensity as experienced with atrazine (ASHTON 1962) showing an action spectrum similar to chlorophylls. In the latter case, triazines would interfere with the electron acceptor of photosystem II. This acceptor termed *Q* (see Fig. 1) (DUYSENS 1963) could react by quenching the fluorescence of chlorophyll (KAUTSKY *et al.* 1960, LAVOREL 1959 and 1962) and would be converted to a substance which may reduce in turn another component of the "cytochrom" chain. By using plastids isolated from the red alga *Porphyricium* sp., NISHIMURA and TAKAMIYA (1966) showed that reduction of $cytochrom_{553}$ by light absorbed by phycobilins was inhibited by simazine. Applying the interpretation of 2 photochemical systems proposed by DUYSENS and AMESZ (1962) and DUYSENS *et al.* (1961) to this type of red algae, the authors suggested that the reduction of $cytochrom_{553}$ by photosystem II was inhibited by simazine, but not its oxidation by photosystem I. From these results and analysis of fluorescence data of chlorophyll in isolated chloroplasts, it was suggested that the simazine as DCMU inhibited one or several reactions close to photosystem II (MURATA *et al.* 1966 a and b). As shown in the "Z"-scheme of electron flow (Fig. 1) developed later, we can now locate the photosystem II further back in the oxidation-

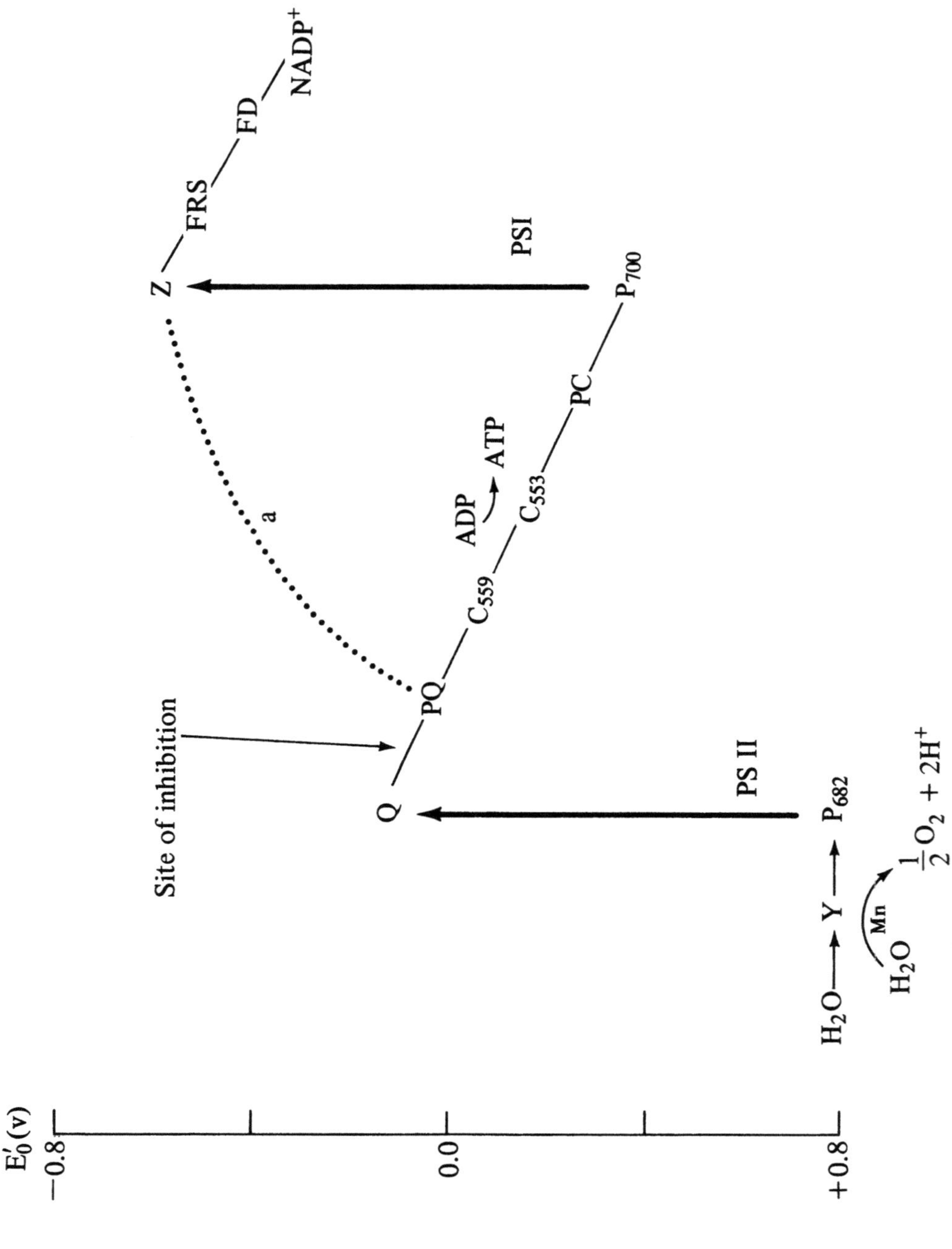

NADP$^+$
FD
FRS
Z
PSI
P_{700}
PC
ATP
ADP
C_{553}
a
C_{559}
PQ
Site of inhibition
Q
PS II
P_{682}
$\frac{1}{2}O_2 + 2H^+$
Mn
H_2O
Y
H_2O
$E_0'(v)$
−0.8
0.0
+0.8

reduction process. The site of action of *s*-triazines could be between water and $cytochrom_{553}$.

From all the experimental data discussed, it is derived that the *s*-triazines, like DCMU, are considered as inhibitors of the acceptor site of photosystem II and would inhibit the reduction of the acceptor of photosystem II at the expense of water oxidation as well as at the expense of oxidation of artificial electron donors for photosystem II (DUYSENS 1963). The exact location of the inhibiting site would be between the primary quencher *Q* of photosystem II and plastoquinone pool.

VAN RENSEN (1971) suggested also a competition between simetone and a substance close to *Q* which accepts the electron back from cyclic electron flow. He proposed a competition between simetone and the oxidized form of this "substance" and *Q*. This was evidenced by a decrease of simetone toxicity with light intensity, some consistent inhibition of cyclic photophosphorylation, a requirement of two electron donors (*Q* and the "substance") during cyclic phosphorylation under white light, and an increase of simetone toxicity in red light compared to that under white light conditions. The substance in question was tentatively identified as plastoquinone, which could react in its oxidized form with simetone by forming hydrogen bonds. However, no direct experimental evidence supported such a hydrogen bonding hypothesis. Other experiments seem to suggest that inhibition or an additional inhibition takes place before photosystem II (ETIENNE 1974, RENGER 1973).

Present findings on possible interactions of atrazine with manganese emphasizes this hypothesis from another point of view. It was shown in physiological experiments that atrazine phytotoxicity was enhanced by phosphorus in cotton and soybeans (ADAMS and ESPINOZA 1969, ADAMS *et al.* 1971). Increase of P-uptake did not increase accumulation of atrazine in soybean green tissues, but it induced a manganese (Mn) deficiency (ADAMS *et al.* 1971, SUN and ADAMS 1971). By increasing the Mn level in the nutrient solution, inhibition of photosynthesis by atrazine could be delayed (SUN and ADAMS 1971). Therefore, it was suggested

---

Fig. 1. Photosynthetic electron flow from water to NADP:
PS I and PS II = photosystems I and II, respectively
Y = natural electron donor of PS II
P_{682} = trap pigment of PS II, a form of chlorophyll absorbing at 682 nm
P_{700} = trap pigment of PS I, a form of chlorophyll absorbing at 700 nm
Q = natural quencher of fluorescence from PS II, primary acceptor of electrons coming from PS II
PQ = plastoquinone
PC = plastocyanin
Z = primary acceptor of PS I, possibly non-heme iron
FRS = ferredoxin reducing substance, function not completely clear
FD = ferredoxin
C_{559} = cytochrom b_{559}, exact location unclear
C_{553} = $cytochrom_{553}$ or cytochrom f
a = cyclic electron flow *in vivo*

that atrazine may block photosynthesis by interacting with the level of chloroplast Mn. The role of Mn in the oxygen evolution system has been discussed and reviewed by GAVALAS and CLARK (1971), CHENIAE and MARTIN (1968), and CHENIAE (1970). A definition of the exact size of the photochemical complex *E* in connection with triazine-inhibited site would be valuable for the understanding of the target of action at the molecular level. RENGER (1970) proposed a model for the natural donor of photosystem II termed *D* or *D II* in which 6 Mn ions were bound. The theoretical value would be 8 Mn/unit of photosystem II, which has been estimated to be around one/250 to 500 chlorophyll molecules (CHENIAE and MARTIN 1968). The site of triazine inhibition would be in one equivalent of an oxygen-evolving unit composed of 250 to 500 chlorophylls and 8 Mn ions. These values are in the same order as the quantasome (BRANTON 1968, PARK and BIGGINS 1964) as suggested by MORELAND (1965) in an oxygen-unit concept, except that the number of Mn ions seems to be different. Triazines may not act only at the Mn level of *Y* because O_2 evolution was reduced in Mn-deficient chloroplasts or algae but not photoreduction (SPENCER and POSSINGHAM 1961). This fact stresses the possible plurality of sites of actions for triazine herbicides.

Also, attempts have been made from physiological studies to explain the phytotoxic action of the triazine herbicides from interference with photosynthesis. The photosynthetic inhibition in *Phaseolus vulgaris* following low and high pre- and posttreatment levels of illumination was followed by VAN OORSCHOT and VAN LEEUWEN (1974). They noted an increased inhibition of net photosynthesis at the lower pretreatment and the higher posttreatment levels of illumination, which the authors ascribe to an enhanced photo-oxidation. ASHTON (1962 and 1965) reported a direct relationship between light intensity and the degree of injury which atrazine and monuron produced in oats and red kidney beans. With increasing amounts of light the toxicity of the herbicide increased. The action spectrum of the phytotoxicity by atrazine fits with the action spectrum of the chlorophylls in photosynthesis, showing a maximum of injury occurring at 428 nm and 658 nm. These results have been interpreted to mean that triazines involve chlorophylls a and b in forming phytotoxic substances which are formed upon illumination. This hypothesis had already been proposed by SWEETSER *et al.* (1961) investigating the mode of action of the phenylureas. However, the phytotoxic intermediates in question have not been characterized.

One of the consequences of the destruction of the photosynthetic apparatus which is visualized by electron microscopy studies (ASHTON *et al.* 1963 b and 1966, HILL *et al.* 1968) is the decrease of the chlorophyll content (BRUINSMA 1965, KALININ and PONOMAREV 1963, MASHTAKOV and PROHORCIK 1962 b, SABUROVA and PETUNOVA 1965, SIVTSEV and KUZNETSOVA 1973, SWIETOCHOWSKI *et al.* 1966) as evidenced by a chlorosis of leaves of triazine-treated plants. Atrazine reduces both the chlorophyll a and the chlorophyll b contents (FREEMAN and RENNEY 1966). No direct interference of triazine with chlorophylls themselves out of their normal chloroplastic site can be demonstrated (IZAWA 1969). A decrease in the carotenoid content was observed and followed the disappearance of chlorophylls (SABUROVA and PETUNOVA 1965, SIVTSEV and KUZNETSOVA 1973). Disappearance of chlorophylls may arise from an irreversible bleaching of the molecules (VAN OVERBEEK 1962). Alternatively, atrazine showed an interaction with magnesium, the uptake of which was reduced by atrazine, and the increase of magnesium level overcame partially the toxicity of atrazine (BRENCHLEY and APPLEBY 1967 and 1971). Therefore, an interference between magnesium and atrazine may be involved in chlorophyll destruction, since magnesium is an essential constituent of this pigment.

The saliant fact in considering the chlorophyll level in triazine-treated plants is its increase, which can precede or follow a previous decrease, and was termed "greening effect" (BARTLEY 1957, COPPING and DAVIS 1972, GAST 1960, KARNATZ 1964, MASHTAKOV and PROHORCIK 1962 b, MOHS 1961, NEWTON 1966, NEWTON and WOODARD 1965, SWIETOCHOWSKI *et al.* 1966, see later chapter on interactions of triazines with plant hormones). This greening effect also occurs with other photosynthesis inhibitors, e.g., *N*-(2-benzothiazolyl)-*N*,*N'*-dimethylurea (FEDTKE 1974).

Chlorophyll and protochlorophyll contents of rice leaves treated with 2,4-dichloro-6-*ortho*-chloroanilino-*s*-triazine were also increased (EGAWA *et al.* 1965, SHISHIYAMA *et al.* 1965). An increased level of chlorophyll and protochlorophyll content may arise from a stimulation of their biosynthesis. In fact, a stimulation of the crude enzyme δ-amino-levulinic acid deshydratase was reported with simazine, IGRAN and GS-14254 (WU *et al.* 1971 b). However, no experiments with the pure enzyme have been carried out. Special attention was paid to the insoluble fraction of the chloroplastic proteins which constitute the lamellar body and the granums, the so-called "holochromic proteins" because of their bonding with pigments. De-novo holochromic protein synthesis seems to be tightly linked with chlorophyll synthesis in the light (MARGULIES 1966, RHODES and YEMM 1963). By incorporating labeled glycine to atrazine-treated shoots of corn which showed a "greening effect," a 2.5-fold increase of the turnover of holochromic proteins was measured (EBERT and VAN ASSCHE 1970). At the same time, no stimulation of chlorophyll biosynthesis from its early precursor was observed, though the chlorophyll content was enhanced to a high degree resulting in visual observation of treated plants greener than the control. The stimulation by triazines of δ-amino-levulinic acid dehydratase seems therefore to be related to a general increase of enzyme biosynthesis rather than to activity. Chloroplast extracted from dark green plants showed a higher specific activity per unit of chlorophyll for reducing ferricyanide. Apparent photosynthesis rates of intact atrazine-treated corn, as measured by CO_2 uptake, were higher than in the control (HIRANPRADIT and FOY 1973, LUND-HOEIE 1969 b,

MASHTAKOV and PROHORCIK 1963 b and 1964, VAN ASSCHE 1971). From these results, one can expect some change in the structure or ultrastructure of chloroplasts. Electron micrographs of chloroplasts extracted from atrazine-treated leaves did not show any change in the ultrastructure. In the meantime, size and weight of these chloroplasts were significantly increased (EBERT and VAN ASSCHE 1970). The amino acid composition of holochromic proteins did not change. Thus the stimulations of apparent photosynthesis would arise from a general increase of the protein content in particular chloroplastic structural proteins which may limit the chlorophyll synthesis.

From all the above-mentioned and in most cases preliminary studies, the beneficial effects of the triazines on photosynthesis remain an interesting tool for theoretical studies.

b) Structure-activity relationships

In the triazine series, simazine was the first which was observed to inhibit the Hill reaction in isolated chloroplasts using Janus green or ferricyanide as electron acceptors (EXER 1958 and 1961, MORELAND *et al.* 1959, MORELAND and HILL 1962). The increasing interest in *s*-triazines as potent herbicides led to the synthesis of a series of compounds which have been tested as Hill reaction inhibitors (VAN ASSCHE 1971). Consequently, the inhibitory activity of chemicals on the Hill reaction has been expressed in terms of molar concentration in the reaction mixture. Their inhibitory potencies were compared to the negative logarithm of the molar concentration causing 50% of inhibition and termed pI_{50}. Values of pI_{50} for a series of 2-chloro-, 2-methylthio- and 2-methoxy-4,6-substituted *s*-triazines are presented in Tables I, II, and III, respectively. A quantitative approach to structure-activity relationships by the use of substituent constants and computerized repression analysis has been made with a series of methylthio-triazines (MOSER 1972).

Some general conclusions can be drawn from the data shown in the different tables. First, when the substitution at the 4-position is the same for 2-chlorotriazines as well as for 2-methylthio- and 2-methoxytriazines, increase in size and/or ramification of the alkyl amino chain increases activity up to a maximum, levels off, and then decreases. This parabolic relationship is not so obvious for 2-methoxytriazines (Tables I-III).

If *Y* and *Z* substitutions are both ethylamino (Table IV), substituents at *X* can be ranked in the following order of inhibition of the Hill reaction: methylmercapto > bromine > methoxy > chlorine > methyl > isopropyl > ethyl > propyl > ethylamino > *n*-butyl. The replacement of Z by a hydroxy group as shown in Table V for some of amino- and alkylaminotriazines causes a drop in the activity of the molecules or divests them of all activity (FOY and BISALPUTRA 1964, VAN ASSCHE 1971). BASKAKOV *et al.* (1965) and GAGARINA *et al.* (1968) compared to the inhibitory activities of several 2-chloro, 4-alkylamino, and 6-alkoxyamino

Table I. ***Inhibition of the Hill reaction by 2-chloro-4,6-substituted s-triazines.***

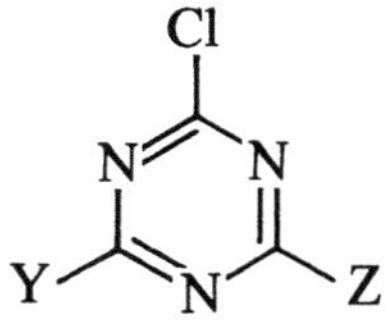

Y	Z	pI_{50}
NH_2	NH_2	0
NH_2	$NHCH_3$	<3
NH_2	NHC_2H_5	3.95
NH_2	NHC_3H_7	4.34
NH_2	NHC_3H_7(iso)	4.65
NH_2	NHC_4H_9	5.82
NH_2	NHC_4H_9(tert.)	5.24
$NHCH_3$	$NHCH_3$	<3
$NHCH_3$	NHC_2H_5	4.46
$NHCH_3$	NHC_3H_7	5.90
$NHCH_3$	NHC_3H_7(iso)	5.05
$NHCH_3$	NHC_4H_9	5.46
$NHCH_3$	NHC_4H_9(tert.)	4.58
NHC_2H_5	NHC_2H_5	5.74
NHC_2H_5	NHC_3H_7	5.40
NHC_2H_5	NHC_3H_7(iso)	5.82
NHC_2H_5	NHC_4H_9	5.29
NHC_2H_5	NHC_4H_9(tert.)	6.45
NHC_2H_5	$NHC(CH_2)_2 - CN$	6.37
NHC_2H_5	$NH(CH_2)_3 - OCH_3$	5.50
NHC_2H_5	NH — H	6.12
NHC_2H_5	$N(C_2H_5)_2$	<3
$N(C_2H_5)_2$	$N(C_2H_5)_2$	3.9
NHC_3H_7(n)	NHC_3H_7(n)	3.52
NHC_3H_7(n)	NHC_4H_9	3.88
NHC_3H_7(n)	NHC_3H_7(iso)	4.30
NHC_3H_7(n)	NHC_4H_9(tert.)	5.22
NHC_3H_7(iso)	NHC_3H_7(iso)	5.67
NHC_3H_7(iso)	NHC_4H_9	5.55
NHC_3H_7(iso)	$NH(CH_2)_2OH$	5.40
NHC_3H_7(iso)	NHC_4H_9(tert.)	5.53
NHC_3H_7(iso)	$NH(CH_2)_3OCH_3$	6.08
NHC_3H_7(iso)	NHC_5H_{11}(iso)	5.82
NHC_3H_7(iso)	$N(C_2H_5)_2$	3.95
NHC_3H_7(iso)	NH —	6.20
NHC_4H_9	NHC_4H_9	<3
NHC_4H_9	NHC_4H_9(tert.)	4.57
NHC_4H_9(sec.)	$NHCH_2 - CH = CH_2$	4.88
NHC_4H_9(tert.)	NHC_4H_9(tert.)	<3
NHC_4H_9(tert.)	NH —	5.92
$NH(CH_2)_3OCH_3$	SCH_3	4.23
$NH(CH_2)_3OCH_3$	NCH_2 o	4.72
$NH(CH_2)_3OCH_3$	$NHCH_2CH = CH_2$	5.28

pI^{50} = the logarithm of the reciprocal molar concentration required to inhibit the reduction of the electron acceptor by 50%.

Table II. *Inhibition of the Hill reaction by 2-methylthio-4,6-substituted s-triazines.*

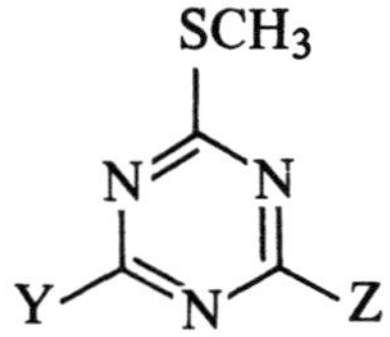

Y	Z	pI_{50}
NH_2	NH_2	<3
NH_2	$NHCH_3$	3.92
NH_2	NHC_2H_5	5.09
NH_2	NHC_3H_7	5.60
NH_2	NHC_3H_7(iso)	5.76
NH_2	NHC_4H_9	5.00
NH_2	NHC_4H_9(tert.)	6.08
$NHCH_3$	$NHCH_3$	5.49
$NHCH_3$	NHC_2H_5	5.85
$NHCH_3$	NHC_3H_7	6.39
$NHCH_3$	NHC_3H_7(iso)	5.85
$NHCH_3$	NHC_3H_9	6.40
$NHCH_3$	NHC_4H_9(tert.)	6.67
NHC_2H_5	NHC_2H_5	6.52
NHC_2H_5	NHC_3H_7	6.60
NHC_2H_5	NHC_3H_7(iso)	6.44
NHC_2H_5	NHC_4H_9	6.95
NHC_2H_5	NHC_4H_9(sec.)	6.85
NHC_2H_5	NHC_4H_9(tert.)	7.15
NHC_2H_5	$NH(CH_2)_3O \cdot CH_3$	6.20
NHC_3H_7	NHC_3H_7	5.82
NHC_3H_7	NHC_3H_7(iso)	6.76
NHC_3H_7	NHC_4H_9	6.08
NHC_3H_7(iso)	NHC_3H_7(iso)	6.67
NHC_3H_7(iso)	NHC_4H_9	6.34
NHC_3H_7	NHC_4H_9(tert.)	6.39
NHC_3H_7(iso)	NHC_4H_9(sec.)	6.52
NHC_3H_7(iso)	NHC_4H_9(tert.)	6.65
NHC_3H_7(iso)	$N(C_2H_5)_2$	4.76
NHC_3H_7(iso)	N_3	3.69
NHC_4H_9	NHC_4H_9	5.42
NHC_4H_9	Cl	<3
NHC_4H_9	NHC_4H_9(tert.)	5.92
NHC_4H_9(tert.)	NHC_4H_9(tert.)	<3

s-triazines on the Hill reaction. In this series, when the substituent at position 2 is chloro, and at position 4 remains the same alkylamino, an increase in size of the alkoxyamino chain decreases activity. The most potent of these was similar in activity to the 2-chloro-4,6-alkylamino-*s*-triazines. However, a comparison between alkoxyamino derivatives and their alkylamino equivalents was difficult because of the lack of common structures. Only ethoxyamino and butoxyamino derivatives were compared to their

Table III. *Inhibition of the Hill reaction by 2-methoxy-4,6-substituted s-triazines.*

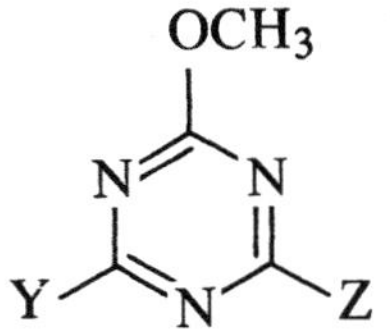

Y	Z	pI_{50}
NH_2	$NHCH_3$	<3
NH_2	NHC_2H_5	3.53
NH_2	$NHCH_2CH = CH_2$	3.56
NH_2	$NH(CH_2)_3OCH_3$	3.76
$NHCH_3$	$NHCH_3$	3.33
$NHCH_3$	NHC_4H_9(sec.)	4.78
$NHCH_3$	NHC_4H_9(tert.)	5.46
$NHCH_3$	NHC_4H_9(iso)	5.11
$NHCH_3$	$NH(CH_2)_3OCH_3$	4.79
NHC_2H_5	NHC_2H_5	4.00
NHC_2H_5	NHC_3H_7(iso)	5.55
NHC_2H_5	$NH(CH_2)_2OCH_3$	4.17
NHC_2H_5	NHC_4H_9(sec.)	5.60
NHC_2H_5	NHC_4H_9(tert.)	6.56
NHC_2H_5	$NH(CH_2)_2OC_2H_5$	4.69
NHC_2H_5	$NHCH - CH_2\text{-}CH(CH_3)_2$ CH_3	6.07
$N(C_2H_5)_2$	$N(C_2H_5)_2$	<3
NHC_3H_7(n)	NHC_3H_7(n)	5.06
NHC_3H_7(iso)	NHC_3H_7(iso)	4.00
NHC_3H_7(iso)	NHC_4H_9(sec.)	5.53
NHC_3H_7(iso)	NHC_4H_9(tert.)	5.53
NHC_3H_7(iso)	NHC_4H_9(iso)	5.45
NHC_3H_7(iso)	$COCH_3$ N C_2H_5	4.37
NHC_3H_7(iso)	$NHCH_2 - CH = CH_2$	4.83
NHC_3H_7(iso)	$NHCH_2CH_2OCH_3$	4.83
NHC_3H_7(iso)	$NHCH_2CH_2CH_2OCH_3$	4.82
NHC_3H_7(iso)	$NHCH_2CH_2OC_2H_3$	4.00
NHC_4H_9(sec.)	NHC_4H_9(sec.)	5.18
NHC_4H_9(sec.)	$NHCH_2CH = CH_2$	4.78
NHC_4H_9(sec.)	$NHCH_2CH_2CH_2OCH_3$	6.69
NHC_4H_9(tert.)	NHC_4H_9(tert.)	5.76
NHC_4H_9(tert.)	$NHCH_2CH_2CH_2OCH_3$	6.61
NHC_4H_9(tert.)	$NHCH_2CH = CH_2$	5.20
NHC_4H_9(iso)	NHC_4H_9(iso)	4.55
$NHCH_2 - CH = CH_2$	$NH(CH_2)_2OCH_3$	4.34
$NHCH_2 - CH = CH_2$	$NH(CH_2)_3OCH_3$	5.61
$NHCH_2 - CH = CH_2$	$NHCH_2 - CH = CH_2$	4.44
$NH(CH_2)_3OCH_3$	$NH(CH_2)_3OCH_3$	5.73

Table IV. *Inhibition of the Hill reaction by bis-ethylamino triazines* (GYSIN and KNUESLI 1960 a).

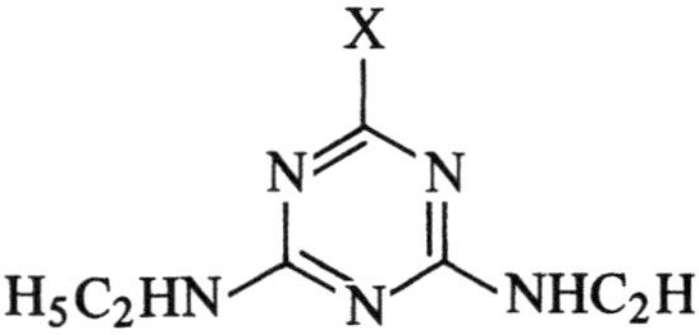

X	pI_{50}
SCH_3	6.83
Br	6.48
OCH_3	6.30
Cl	6.13
CH_3	6.12
$CH(CH_3)_2$	5.52
C_2H_5	5.46
$n - C_3H_7$	5.33
NHC_2H_5	5.05
C_4H_9	3.79
H	3.00

Table V. *Inhibition of the Hill reaction by 2-hydroxy-4,6-amino and alkylamino s-triazines.*

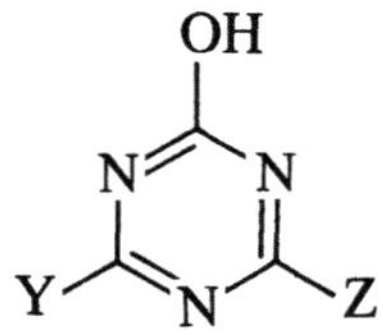

Y	Z	pI_{50}
NH_2	NH_2	0
NH_2	NHC_2H_5	0
NH_2	NHC_3H_7(iso)	0
NH_2	NHC_4H_9(sec.)	0
NH_2	NHC_4H_9(tert.)	0
NHC_2H_5	NHC_4H_9(sec.)	<3
NHC_2H_5	NHC_4H_9(tert.)	3.79
NHC_2H_5	NHC_3H_7(iso)	<3
NHC_3H_7(iso)	NHC_3H_7(iso)	<3

alkylamino analogues; they showed a lower activity. More recently, another series of 2-azido-4,6-bis-alkylamino triazines has shown that the replacement of a Cl in position 2 by the azido group does not change significantly the rank of inhibitory potencies as presented in Table VI. A general conclusion can be drawn from the results obtained with either 2-chloro or 2-azido bis-alkylamino triazines: an increase in activity occurs

Table VI. *Inhibition of the Hill reaction by 2-azido-4,6-alkylamino-s-triazines* (GABBOTT and BARNSLEY 1968).

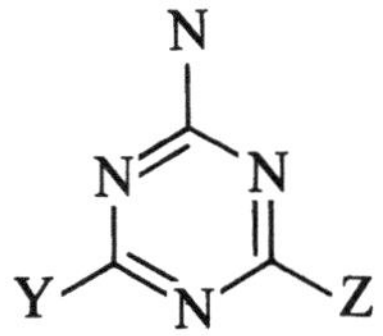

Y	Z	pI_{50}
$NHCH_3$	NHC_2H_5	5.26
$NHCH_3$	NHC_3H_7(*n*)	5.50
$NHCH_3$	NHC_3H_7(iso)	5.82
$NHCH_3$	NHC_4H_9(*n*)	6.14
$NHCH_3$	NHC_4H_9(iso)	6.11
$NHCH_3$	NHC_4H_9(sec.)	6.49
$NHCH_3$	NHC_4H_9(tert.)	6.75
$NHCH_3$	NHC_3H_{11}(tert.)	6.70
NHC_2H_5	NHC_2H_5	5.99
NHC_2H_5	NHC_3H_7(*n*)	6.13
NHC_2H_5	NHC_3H_7(iso)	6.54
NHC_2H_5	NHC_4H_9(*n*)	6.80
NHC_2H_5	NHC_4H_9(iso)	6.64
NHC_2H_5	NHC_4H_9(sec.)	7.04
NHC_2H_5	NHC_4H_9(tert.)	7.46
NHC_2H_5	NHC_5H_{11}(tert.)	6.77
NHC_3H_7(iso)	NHC_3H_7(iso)	6.87
NHC_3H_7(iso)	NHC_4H_9(tert.)	6.85

in a parabolic fashion with increasing chain length. Maximum activity is noticed with compounds having mixed substitutions possessing a total of 5 to 6 alkyl carbon atoms.

Attempts have been made to find a correlation between the inhibition data and the physico-chemical properties of substituents of a given molecule. Some suggestions were offered by MORELAND and HILL (1962) and HILL (1965) such as the following:

(1) A favorable balance between lipophilic and hydrophilic groups may enable the chemical to pass through or allow partition into the lipoprotein matrices of the chloroplast to reach the reactive site(s).

(2) A particular spatial configuration can have a permissive role for covering or fitting into the reactive site(s).

(3) The orientation of the 2-chloro or 2-alkoxy triazine molecule in the active site(s) may be controlled by steric factors of alkylamino substituents in positions 4 and 6.

(4) No correlation can be found between the water solubility of triazine derivatives and inhibition of Hill reaction.

These hypotheses have been otherwise formulated by using extra thermodynamic approach (Hansch approach), first employed with benzoic acid derivatives, substituted phenyl amides, ureas (FUJITA *et al.* 1964, HANSCH *et al.* 1965, HANSCH 1969, HANSCH and FUJITA 1964). Accordingly, a model was proposed by using a combination of free energy-related substituent constants and those of steric and partition effects to make a multiple regression analysis. Such an equation was not developed for the 2-chloro-4,6-alkylamino triazines, but it was only recently introduced for 2-azido-4,6-alkylamino triazines (GABBOTT 1968).

The multiple regression analysis utilizing the above-mentioned constants gave values of calculated inhibitory potencies which correlated very well to the pI_{50} determined experimentally. This approach proved to be successful for many important herbicide classes as benzimidazoles, phenylcarbamates, anilides, ureas, and triazines (BUECHEL 1972, HANSCH 1969). The most striking observations were the increasing effects from additions to the alkyl substituents to a point and then the parabolic nature of the data with further increases to the alkyl chain length. These effects could be envisaged in terms of the affinity of the inhibitor for the target site and/or reactivity at the active site(s) (GABBOTT 1969, HANSCH and DEUTSCH 1966). Research of the same type based on partition data for a series of 2-alkylthio-4,6-bisalkyl-amino triazines has been performed (MOSER 1972). However, the remarkable correlation experienced with azido triazines did not appear so significantly with alkylthio triazines. Since most of the tested compounds were herbicidally active triazines, the range of pI_{50} was not so widely spread; moveover, the standard error of the pI_{50} values was sometimes of the same magnitude as differences in inhibitor potencies.

All the photosynthetic active triazines have as a basic chemical structural element a -C̈-NH- group. This structural element is a general requirement for the inhibitors of photosystem II like the ureas, anilides, uracils, etc. The lipophilic substituent at the nitrogen is usually an aromatic ring (HANSCH 1969, MORELAND 1969, BUECHEL 1972). Complete *N*-alkylation of the triazines and the above-mentioned chemical classes lead to inactive compounds (MORELAND 1969). One known exception with active triazines is the diethyl derivative of simazine, chlorazine. It has a pI_{50} of $1.2 \times 10^{-4} M$ which is not very active as a Hill inhibitor, but, nevertheless, it has very distinct herbicidal activities.

It was suggested that the herbicidal element -C̈-NH- might form hydrogen bonds between the amide hydrogen and the active centers of the chloroplast, e.g., peptide bonds in enzymes (BARTH and MICHEL 1968 and 1969; CAMPER and MORELAND 1965; GOOD 1961; MORELAND 1967; MORELAND and HILL 1959, 1962, and 1963; VAN OVERBEEK 1962; WESSELS and VAN DER VEEN 1956). Particularly HANSCH (1969) stressed this point. But because exceptions have been found, the hydrogen bond theory was questioned. The inhibitor potencies of anilides were not necessarily correlated with the hydrogen bonding data. Furthermore, the amino hydrogen(s) of the triazines did not bind to the carbonyl oxygen of acetone to an appreciable extent (GOOD 1961). BRIAN (1965) pointed out that the use of acetone as a simulated biological model does not allow a comparison of these results to the bonding with an active site of the chloroplast. Therefore, already HANSCH (1969) and in particular BUECHEL (1972), supposed that a common requirement for inhibitory activity would consist of a planar arrangement of atoms in which a nitrogen atom is attached to an electron deficient sp_2C atom. In the proposed model the single pair electrons of nitrogen would interact with the electron-deficient carbon of a protein carbonyl group, while the electrons of the nitrogen atom of protein amide linkage could interact with some electron-deficient atom of the inhibitor. From this follows that charge transfer and dipole interchanges are more important than hydrogen bonding. Recently, particularly striking examples of highly active *N*-alkylated ureas have been reported, which are photosystem II inhibitors (MORELAND 1969, TREBST and HARTH 1974). A revised basic element was postulated to include also heterocyclic compounds, active as photosynthesis inhibitors (TREBST and HARTH 1974).

It should be pointed out that the NH group has been widely termed by numerous authors "imino" for phenylureas, phenylamides, and carbamate herbicides. In the case of the alkylamino triazines, this designation was also used and might lead to some misinterpretation (BRIAN 1965). In fact, either the NH of alkylamino chain must be termed amino (IUPAC rules), or the term "imino" was applied to the ring nitrogen of the resonance form, which is unlikely.

In comparative experimental studies, the determination of the exact concentration of the active molecule plays a major role. In order to estimate the exact concentration of the inhibitor in the chloroplasts, IZAWA and GOOD (1965) used two methods. The first one, called "chloroplast-amount-method," consisted of an increase in the concentration of chloroplasts until the same degree of inhibition was obtained, assuming that this degree was dependent on the amount of chloroplast present. In the second method, or "chloroplast-removal-method," the chloroplasts were centrifuged and the remaining inhibitor concentration was estimated in the supernatant. They concluded that absorption of atrazine by the thylakoid membrane involved at least three simultaneous processes:

(1) an irreversible binding of the inhibitor not associated with inhibitory activity in a ratio of about one molecule per 1,000 chlorophyll molecules,

(2) a partition between the aqueous phase and the chloroplasts (biological phase) independent of inhibitor concentration and the inhibitory range, and

(3) an absorption of the chemical to the thylakoid membrane which corresponds closely to the degree of inhibition.

The number of sites of inhibition seemed to be one for every 2,500 chlorophyll molecules (GOOD and IZAWA 1964, IZAWA and GOOD 1965). This estimation is important, since it may correspond to one oxygen-evolving unit, as will be discussed later. Other authors suggested that this ratio was nearer one per few hundred molecules of chlorophyll (MONACO 1968, MONACO and MORELAND 1966). This apparent discrepancy was discussed by CHENIAE and MARTIN (1968), who measured the quantum requirement for O_2 evolution and concluded that not all the sites were active and consequently the ratio 1:2,500 might be overestimated.

STROTMANN *et al.* (1974) introduced a more refined method for measuring the inhibitory constant. Using radioactive labeled inhibitors, they measured the free inhibitor concentration and calculated from this the concentration of bound inhibitor. From this they concluded:

(1) The half-saturating concentrations for the inhibition of the electron transport by atrazine and its absorption by the thylakoid membrane correspond to each other ($I_{50} = 0.51\ \mu M$, $A_{50} = 0.47\ \mu M$). Both concentration curves are rectangular hyperbolae. There is no unspecific cohesion of atrazine by the thylakoid membrane; the cohesion is exclusively due to the electron-transport component which is sensitive to atrazine.

(2) The maximum cohesion of atrazine by the thylakoid membrane is 4.7 nMol/mg of chlorophyll. Supposing that at a time one molecule of the electron transmitter to be inhibited by atrazine complexes with one molecule atrazine, we come to the ratio of 1 Mol of the electron transmitter to 430 Mol of chlorophyll. This value corresponds well to the concentrations given in the literature for P 700, cytochrom f, ferredoxin, and plastocyanin.

(3) The cohesion of atrazine to the thylakoid membrane is competitively inhibited by DCMU and biscarbamate. This results in the fact that all three types of inhibitors inhibit the same component of the photosynthetic electron transport.

The Hill activity of atrazine- or simetone-inhibited chloroplast can be restored completely by washing the chloroplast (MORELAND and HILL 1962, VAN RENSEN 1971, VAN RENSEN and VAN STEEKELENBURG 1965).

Several remarks have to be made on the results of the structure-activity studies. First, herbicidal activity does not necessarily parallel the Hill reaction inhibitor potencies of chemicals (BASKAKOV 1965, GABBOTT and BARNSLEY 1968, GYSIN and KNUESLI 1960 a and b). Too many factors may be involved, such as adsorption by the soil, translocation, solubility, metabolism of the chemical, etc. Nevertheless, inhibition of the Hill reaction seems to be essential for herbicidal properties of *s*-triazines. Although the rank of inhibitory potencies is almost the same in the various publications, the absolute value of pI_{50} differs to some extent. This may be due to differences in techniques for separation of chloroplasts, origin of the plant material, or assay conditions for the Hill reaction. For instance, EXER (1961) pointed out that the use of Janus green as an electron mediator seemed to increase the apparent toxicity of triazines, leading to overestimated values of pI_{50}. Also, water solubilities of some *s*-triazines are very critical (GUNTHER *et al.* 1968, GYSIN and KNUESLI 1960 b). In some cases, inhibition curves showed the normal sigmoid pattern, even at concentrations far from the limit of solubility of the compound; consequently, even pI_{50} values were higher than the concentration of the chemical theoretically soluble in water. Herbicides were shown to adsorb onto proteins so that active concentration in the medium would be changed and produce a shift in relative inhibitory potencies (CAMPER and MORELAND 1971). The relative liposoluble character of some triazines may enable them to be adsorbed readily by lipid membranes or lipoproteins. The structure-activity relationships have shown that lipophilie of the compound involved is necessary for their binding to membranes. To be effective, the herdicide must also have some water solubility in order to be taken up by the roots and distributed in the xylem through the plant.

The target site of triazines within the photosynthesis process remains only partially understood because of the lack of knowledge of the primary events of light reactions at the molecular level. Progress made recently on the biochemistry of photosynthesis together with an improvement in analytical techniques should lead rapidly to a better understanding of the intimate events responsible for water-splitting phenomena. Inhibitors of various structures possessing herbicidal activity would widely help to identify the mechanism involved. In turn, such new data will permit a better comprehension of the exact mode of action of herbicides including triazines.

II. Respiration

Several reports deal with the effects of the triazines on the oxidative respiratory system of intact plant tissues (AGHA and ANDERSON 1967, ALLEN and PALMER 1963, ASHTON 1960, BUSCHBECK and KRISPIN 1971, CHODOVA and ZEMANEK 1971, DALVI *et al.* 1972, FUNDERBURK and DAVIS 1963, HARTISCH 1967, HARTMANN 1971, KOZLOWSKI and SASAKI 1966, KRIHNING 1965, MASHTAKOV and PROHORCIK 1964 and 1966, MASHTAKOV *et al.* 1969, NASYROVA *et al.* 1968, OLECH 1967 and 1968, RIES *et al.* 1967 a, ROTH 1959, SASAKI and KOZLOWSKI 1966 and 1969 a and b, SINZAR 1967, SPIESIVTSEV *et al.* 1969 a and b, VELEV 1969, and VOINILO *et al.* 1967). RIES *et al.* (1967 a) showed that rye plants grown in 2 to 80 $\times$ $10^{-8}M$ simazine nutrient solution increased their respiration by 10% without affecting the respiratory quotient. Similar observations were made in the earlier work of ROTH (1959) with *Elodea* and of ASHTON (1960) with excised bean leaves. The study by ALLEN and PALMER (1963) showed that the respiration of excised barley roots was not influenced by simazine alone, but the addition of glucose, aspargine, aspartate, or α-ketoglutarate along with simazine produced various effects. These effects ranged so widely that no definite conclusion could be drawn for the role of simazine. In most studies mentioned above, the respiration was decreased or left unchanged by various triazines when applied at herbicidal rates to plants such as corn, peanuts, red pine, barley, grain sorghum, wheat, oats, tomatoes, cotton, and soybeans. Where the triazines were reported to stimulate respiration, the experiments were only of a few hours duration. In longer term experiments, the triazines tended to retard respiration in both resident and susceptible species (FUNDERBURK and DAVIS 1963). In all cases where photosynthesis activity was also reported, it was inhibited with a factor of 100 more severely than respiration. This would indicate that the inhibition of oxidative respiration in whole plants cannot play a major role in the herbicidal action of the triazines. Indeed, OLECH (1968) claimed that atrazine affects respiration indirectly by inhibiting photosynthesis, which causes a reduction in assimilated products.

When isolated mitochondria from plant tissues such as beans, cabbage, cucumbers, corn, soybeans, and peas and even from animal tissue (rat liver) are treated with a number of triazines at concentrations ranging from 10^{-3} to $10^{-5}M$, some inhibition of oxygen uptake occurs (DAVIS 1968, FOY and PENNER 1965, LOTTIKAR *et al.* 1968, MCDANIEL and FRANS 1969, PARKS *et al.* 1972, THOMPSON 1973, THOMPSON and DAVIS 1969, THOMPSON and TRUELOVE 1972, THOMPSON *et al.* 1969 and 1974, TRUELOVE and DAVIS 1969, VOINILO *et al.* 1967 and 1969). The methoxy- and methylthio-triazines are more effective in this regard than the chlorotriazines although many investigations have been made with the latter. FOY and PENNER (1965) found that atrazine at 3×10^{-4} to $3 \times 10^{-5}M$ inhibited oxygen uptake by cucumber mitochondria, but lower concentrations ($3 \times 10^{-6}M$) produced a light stimulation. However, LOTTIKAR *et al.* (1968) found only a slight reduction in oxygen uptake and P:O ratios in cabbage mitochondria in the presence of $6 \times 10^{-5}M$ simazine, but this may not have been significant. DAVIS (1968) reported that atrazine at concentrations as high as $4 \times 10^{-4}M$ has no effect on corn mitochondrial respiration or on the P:O ratios. The ATP-level of soybean hypocotyls was not affected by $2 \times 10^{-4}M$ atrazine. Also, HACHIMORI *et al.* (1968) found no effect of $1 \times 10^{-4}M$ simazine or atrazine on yeast mitochondria. All the evidence indicates that the chlorotriazines have a minor effect on mitochondria, but a definite inhibition of mitochondrial respiration occurs with amethryn and prometryn. In the respiratory chain, the state 3 respiration is inhibited, which is the stage where the respiratory chain is coupled to the ATP synthesis (MCDANIEL and FRANS 1969, PARKS *et al.* 1972, THOMPSON *et al.* 1969 and 1974). Stage 4 respiration was not affected by the triazines (THOMPSON *et al.* 1974).

In order to discuss the possible site of action of the triazines on mitochondrial respiration, the following scheme is offered which makes use of the terminology proposed by CHANCE and WILLIAMS (1956): during respiration, a substrate is oxidized and the resulting electrons are transferred to molecular oxygen, reducing it to water. The energy from these oxidation-reduction reactions is trapped by the hypothetic intermediate carriers $X + I$. The result is the formation of a high-energy intermediate

$X \sim I$, which can break down spontaneously. With no energy acceptor in the system, *i.e.*, inorganic phosphate (P_i) and adenosine diphosphate (ADP), spontaneous breakdown is the only way for the energy in the high-energy-intermediate complex to escape (known as state 4 inhibition). With the P_i and ADP added to the system, another path for dissipating the $X \sim I$ is opened. The energy in $X \sim$ I is trapped in adenosine triphosphate (ATP) through the action of mitochondrial ATPase, and the respiration proceeds at a faster rate known as the state 3 inhibition. Uncouplers such as dinitrophenol or carbonyl cyanide *m*-chlorophenylhydrazine stimulate the spontaneous breakdown of $X \sim I$ and have been shown to release the state 3 inhibition caused by triazines (McDaniel 1969, Thompson *et al.* 1969 and 1974). From the results the conclusion was drawn that ATP-synthesis might be a site of action of several triazines on mitochondria. Inhibition of electron transport in the mitochondria as a site of action is excluded, since the uncoupled and state 4 respiration are not affected by them (Thompson *et al.* 1974).

A survey on all the data from respiration studies with triazines in plants, including physiological and biochemical data, has shown that the effects of triazines on mitochondria in comparison to those in chloroplasts are of minor importance in the whole spectrum of events of the triazines on plant physiology.

III. Plant constituents

a) Dry matter and protein production

There would be no purpose for discussing the changes in plant constituents of a dying plant or one suffering from toxic symptoms caused by herdicides, but changes caused by subherbicidal levels of the *s*-triazine herbicides have been enthusiastically reported for over a decade. Much excitement centered around the possibility of using low levels of these herbicides to increase both the yield and the protein content in many crops and, indeed, several different patents have even been granted for this use (*Asahi Chem. Ind. Co. Ltd.* 1967, *Battelle Dev. Corp.* 1968, *Standard Oil Co.* 1968, *Nippon Kayaku KK* 1969, and *Moldav Seed Select Inst.* 1972). In spite of nearly a dozen years of research into the reasons for these subtle changes in plant composition from sublethal additions of *s*-triazines, we still can only speculate about the nature of why they occur. Some of the more likely explanations involve changes in the plants energy balance, changes in the water balance, or effects on hormones and enzymes which are discussed in later sections.

Of all the nonherbicidal effects so far noted for the *s*-triazines, none has been reported as profusely as the effect on nitrogen metabolism. Several hundred references exist on this one topic alone, and there are reports on dry matter and nitrogen increases in a variety of crop plants (see Table VII for a representative listing of the more pertinent literature). An

Table VII. *Influence of triazines on nitrogen and protein contents in plants.*

Plant species	Latin name	Triazine involved	Reference
Barley	*Hordeum vulgare* L.	simazine	ALABUSHEV (1965)
		simazine	BASTIN *et al.* (1968 and 1970)
		simazine	PULVER & RIES (1973)
		simazine	RIES & WERT (1972 a and b)
Corn	*Zea mays* L.	simazine	ALABUSHEV (1965)
		atrazine	BAGAEV & KOBAZEVA (1965)
		simazine	COIC & LESAINT (1971)
		atrazine	DOMANSKA (1962)
		simazine, atrazine	FINK (1967)
		simazine, atrazine	FINK & FLETCHALL (1967)
		simazine, atrazine	FOLAND (1966)
		simazine	FRENEY (1965)
		simazine	FRIESKE (1971)
		atrazine	GAST & GROB (1960)
		atrazine	GRAMLICH *et al.* (1965)
		atrazine	GRAMLICH & DAVIS (1967)
		simazine	GZIK & GRAESER (1972 a)
		atrazine	HIRANDAPIT *et al.* (1972)
		atrazine	LAY & ILNICKI (1972)
		simazine	MAZUR & KAWECKA (1969)
		simazine, atrazine	PULVER & TWEEDY (1968)
		simazine	RIES & GAST (1965)
		simazine, chlorazine, atrazine	SUSU (1966)
(Grain)		simazine, chlorazine, atrazine	SUSU *et al.* (1970)
		simazine	TWEEDY & RIES (1967 a and b)

(Grain)		simazine, atrazine	Volodin (1968)
		simazine	De Vries (1963)
Sweet corn (seed)	*Zea mays* L., *cv. lochief*	simazine, atrazine, prometon, IGRAN, ametryn	Salunkhe *et al.* (1971 a)
		simazine, atrazine, prometon, IGRAN, ametryn	Salunkhe *et al.* (1971 b)
		simazine, atrazine, prometon, IGRAN, ametryn, GS-14254, propazine	Wu *et al.* (1971 a and c and 1972 a and b)
		atrazine	Greig & Asif (1971)
Various grasses	*Festuca arundinacea Schreb., Phleum pratense* L., *Lolium perenne* L., *Lolium multiflorum Lam., Phalaris arundinacea* L.	simazine	Allison & Peters (1970)
Barnyardgrass	*Echinachloa crus-galli* (L.) *Beauv.*	simazine	Chmiel & Ries (1967)
Wheatgrass	*Agropyron intermedium* (*Host.*) *Beauv.*	atrazine	Kay (1971)
Red brome	*Bromus rubens* L.	atrazine, simazine	Kay (1971)
Arabiangrass	*Schismus arabicus Nees*	atrazine, simazine	Kay (1971)
Red-stem filaree	*Erodium circutarium* (L.) *L'Herit.*	atrazine, simazine	Kay (1971)
Bermudagrass	*Cynodon dactylon* (L.) *Pers.*	simazine	Monson *et al.* (1971)
Ryegrass forage	*Lolium perenne* L.	simazine	Ries *et al.* (1968)
Blue gramagrass	*Bouteloua gracilis* (*H.B.K.*) *Lag.*	atrazine	Tapia (1974)
Johnsongrass	*Sorghum halepense* (L.) *Pers.*	atrazine	Gramlich (1965)
Quackgrass		simazine	Swietochowski *et al.* (1968)
Oat	*Avena sativa* L.	atrazine	Eastin & Davis (1967)
		prometryn	Iwanowa & Tschorbadshijska (1965)
		simazine	Mazur & Kawecka (1969)

Plant species	Latin name	Triazine involved	Reference
		simazine	Ries *et al.* (1969)
		simazine	Ries & Wert (1972 a and b)
		simazine	Saburova & Petunova (1965)
		simazine	Singh (1964)
Oat (seed)		simazine	Schweizer & Ries (1969)
Rice (seed)	*Oryza sativa* L.	simetryn	De Datta *et al.* (1972)
		simazine	Vergara *et al.* (1970)
		simetryn	*International Rice Research Institute* (1971)
		simazine	Miller & Mikkelsen (1970)
Rice		simazine	Ries *et al.* (1967 a)
Rye	*Secale cereale* L.	atrazine	Budoi & Ries (1968)
		simazine	Chmiel & Ries (1967)
		simazine	Ploszynski (1973 and 1974)
		simazine	Ries & Wert (1972 a and b)
Rye (seed)		simazine	Ries *et al.* (1967 a)
Sorghum	*Sorghum* sp.	atrazine	Gramlich *et al.* (1965)
		atrazine	Gramlich & Davis (1967)
		atrazine, simazine	Tweedy *et al.* (1971)
Wheat	*Triticum aestivum* L.	IGRAN, simazine	Bastin *et al.* (1970)
		simazine	Brunetti *et al.* (1971/72)
		simazine	Schweizer & Ries (1969 a)
		simazine, atrazine	Tweedy *et al.* (1971)
		prometryn	Lalova (1969)
Wheat (seed)		simazine, atrazine	Konovalova (1967)
		simazine	McNeal *et al.* (1969)
		simazine	Ries *et al.* (1970)

Trees:			
Apple	*Malus sylvestris Mill.*	simazine, atrazine	LIN & ANDERSON (1967)
		simazine, atrazine	KARNATZ (1964 and 1965)
		simazine	LIN (1967)
Apple & peach	*Prunus persica Batsch.*	simazine	RIES *et al.* (1963)
Apple		simazine	SOLECKA *et al.* (1969)
Citrus	*Citrus sinensis* (L.) *Pers.*	simazine, atrazine	GOREN & MONSELISE (1966)
Oak	*Quercus* sp.	simazine, prometryn	SHIMANSKII (1970)
Pine	*Pinus* sp.	simazine, atrazine	CONNER & WHITE (1968)
		simazine, atrazine	CONNER (1969)
		simazine	IVANOVA (1966)
		simazine, prometryn	SHIMANSKII (1970 and 1971 a)
		simazine	DE VRIES (1963)
Varia			
Alfalfa forage	*Medicago sativa* L.	simazine	RIES *et al.* (1968)
Beans	*Phaseolus vulgaris* L.	atrazine	EASTIN & DAVIS (1967)
		atrazine, simazine, terbutryn, GS-14254	SINGH *et al.* (1972)
Peas	*Pisum sativum* L.	simazine	ALABUSHEV (1965)
		simazine, atraton	PLOSZYNSKI (1972)
		prometryn	SCHULKE (1970)
		simazine	VOROB'EV & CH'A (1960)
Beans & peas (seed)		simazine	RIES *et al.* (1967 a)
		simazine	RIES *et al.* (1968)
Peas (seed)		simazine, propazine, IGRAN, ametryn	SALUNKHE *et al.* (1971 a and b)
Cotton	*Gossypium hirsutum* L.	atrazine	EASTIN & DAVIS (1967)
		ipazine, prometryn	NASYROVA *et al.* (1968)
Grape-vine	*Vitis vinifera* L.	simazine	TARLAPAN & PANCYR (1966)

Plant species	Latin name	Triazine involved	Reference
Lupin (seed)	*Lupinus luteus* L.	simazine, atrazine	DOROSHENKO & YATSYUK (1968)
Peanut	*Arachis hypogae* L.	atrazine	EASTIN & DAVIS (1967)
Potatoes	*Solanum tuberosum* L.	simazine	GARUS *et al.* (1963), VIGUE & LI (1974)
		simazine	MAZUR & KAWECKA (1969)
		atrazine, prometryn	SVRKATA *et al.* (1968)
Radishes	*Raphanus sativus* L.	simazine	GNANARETHINAM & CARLES (1966)
		simazine	MAZUR & KAWECKA (1969)
Soybeans	*Glycine Max* (L.) *Merr.*	atrazine	EASTIN & DAVIS (1967)
		atrazine	GRAMLICH (1965)
Sunflower	*Helianthus annuus* L.	atrazine	CHESALIN & TINOFEEVA (1968 and 1971)

attempt has been made to select the important articles needed to discuss adequately all ramifications of this interesting effect.

After BARTLEY (1957) reported that simazine-treated corn was greener and yielded more than hand-weeded controls, many investigations in corn and in many other crops have been made with enthusiasm for the possible extra economic value of using such a herbicide. GAST and GROB (1960) reported that normal preemergence applications of simazine or atrazine could increase the yield of forage corn by 2 to 3 times that of untreated plots. They also noted increases in dry weight, crude protein, and carbohydrate content in the corn tissue. DE VRIES (1963), on the other hand, found that simazine would reduce the dry weight of the roots and tops of corn and *Pinus radiata*, but he did find increased uptake of nitrogen from all soils and of magnesium and phosphorus in limed and of potassium in unlimed soils.

Influences of triazines on potassium and/or phosphorus uptake into plants are reported by BAGAEV and KOBAZEWA (1965), BARANOWSKI *et al.* (1967), CONNER (1969), DHILLON *et al.* (1967), GILL *et al.* (1970), KALIMULLINA and BAEGEV (1967), LEHMANN *et al.* (1971), MILLIKAN *et al.* (1966), PLOSZYNSKI (1973), SHIMANSKII (1970 and 1971 a and b), SOSNOVAYA (1971), ZINCHENKO *et al.* (1966), and ZHIRMUNSKAYA and JOFFE (1973).

SMITH and MCVEIGH (1971) saw that phosphorus, potassium, and magnesium levels were increased but not calcium from atrazine treatments. They thought that this was a result of atrazine stimulating some energy-producing process which drives active ion transport, because calcium was considered as being transported passively while the others were transported actively.

ZHIRMUNSKAYA and JOFFE (1973) studied the effect of atrazine on potassium absorption by barley roots. Atrazine (1 to 14 mg/L) in nutrient solutions inhibited the absorption of potassium. In a previous paper, ZHIRMUNSKAYA and KOLTZOVA (1973) had shown that the uptake of atrazine from the nutrient solution into barley roots was only due to adsorption and that the surface cell membranes of barley roots possess a high permeability toward atrazine. They explain the inhibition of potassium uptake by atrazine to arise from a reversible adsorption of atrazine on the surface of the root cells.

RIES and his associates have reported a large number of protein responses of various crops to sublethal concentrations of the triazines (CHMIEL and RIES 1967, RIES *et al.* 1967 a and b, RIES and GAST 1965, RIES *et al.* 1968 and 1970, SCHWEIZER and RIES 1969 a and b, TWEEDY and RIES 1967 a and b). However, many reports show either no effect, or the extra responses are offset by a reduction in dry weight of the plant (MONSON *et al.* 1971, TWEEDY *et al.* 1971, VERGARA *et al.* 1970). Many of the nitrogen-protein responses only occur when nitrogen is low or below optimum in the growing medium (TWEEDY and RIES 1967 a and b, TWEEDY *et al.* 1971, LAY and ILNICKI 1972, and PLOSZYNSKI 1973).

Tweedy and Ries (1967 a and b) and Doll and Meggitt (1968) found that nitrogen content would increase in corn after triazine treatment only when the temperature was low, and furthermore, when nitrate is replaced by the ammonium ion in nutrient cultures, no extra stimulation of nitrogen uptake occurs. Whatever the mechanism for triazines increasing plant nitrogen may be, it seems sure to involve the nitrate ion uptake mechanism.

Since the first visual response which Bartley (1957) and others noted was a greening similar to a nitrogen fertilizer response, it was only natural to investigate the total nitrogen content of simazine-treated plants. When several data showed that extra protein-nitrogen was present in many treated plants, it was thought to be an effect on the nitrification process. Freney (1965) could find no effect of simazine on mineralization of soil organic nitrogen and suggested that simazine increased plant growth by a direct effect on plant metabolism and not by any interaction with the soil. Likewise, Grossbard (1970) found that the effect on soil mineral nitrogen from repeated applications of simazine over a five-year period was not significant even though there was a small depression in the soils CO_2 output. If the bacteria responsible for mineralization, nitrification, or nitrogen fixation were stimulated, then a ready explanation would exist for the extra nitrogen found in some triazine-treated plants. This entire area was reviewed by Kaiser *et al.* (1970), and they found many conflicting reports. Several authors have reported increases in nitrogen fixation and even nitrification, but nearly as many have reported no effect or slight retardations (Caseley and Luckwill 1965, Eno 1962). Undoubtedly the environmental conditions play a very important role in determining how bacteria respond to herbicides. Sensitive species are affected more in poor light soils and by strong doses of herbicides which might explain drops in nitrogen uptake. But, so far, no explanation can be given for the stimulations noted. Although some reports show that certain soil microbes can utilize the nitrogen of atrazine or nitrify it once it is released (Hauck and Stephenson 1964, Kaiser and Reber 1970), it cannot be of any consequence in making extra nitrogen available for plant uptake.

Several authors investigated the influence of the triazines on nitrogen fixation in leguminosae by *Rhizobium* sp. bacteria. From their results in greenhouse and field experiments, they agree that concentrations of triazine herbicides which reduce plant height and dry matter content also reduce the number, weight, and activity of the nodules, (Hauke-Pacewiczowa 1970, Schulke 1970, Thomas and Hammond 1968, Vagina and Latypova 1972, Velev 1971). Elenkov and co-workers (1970) found in field experiments that 2 kg/ha prometryn increased the number of nodules in French beans. However, they could not decide whether this effect was beneficial action of the prometryn on plant growth or a direct effect on nodule formulation. Kaszubiak (1966) investigated the sensitivity of *Rhizobium* sp. to a series of herbicides in a Thornton liquid medium with yeast. Simazine and prometryn were not toxic to the *Rhizobium* strains tested. SCHULKE (1970) compared in more detail the growth behavior of pea plants with and without nodules under the influence of prometryn. He found a growth-stimulating effect of the chemical on the pea plants without nodules which were kept in a nutrient solution containing nitrogen. There was no growth stimulation in nodule containing plants which were kept in a nutrient solution without nitrogen. Increases in the number of nodules were accompanied by decreases in nodule activity. Therefore, it seems unlikely that nitrogen increases in triazine-treated plants can be explained by stimulation of nitrogen fixation. There are several reports where increased plant nitrogen occurred without any changes in the activity of these processes (FRENEY 1965 and IVANOVA 1967). Until such time as correlation studies prove positive, we must consider that factors other than those in the soil are responsible for the plant growth stimulations produced by the triazines. Indeed several studies were carried out with plants growing in nutrient solutions (BASTIN *et al.* 1970, RIES and GAST 1965, TWEEDY and RIES 1967 a and b), which further discount biological soil factors as being important as far as contributing extra nitrogen for plant uptake. Nevertheless, physico-chemical soil factors may contribute to plant responses to triazines if we consider that factors such as temperature, acidity, moisture, base saturation, and presence of added fertilizers all affect the amount of triazine taken into the plant. MCREYNOLDS and TWEEDY (1970) found that rye, corn, and oats took up more ^{14}C-labeled simazine from nutrient solutions containing nitrate nitrogen than ones containing only ammonium nitrogen. Oats were injured less on soils fertilized with calcium nitrate. The physico-chemical soil factors may even be the reason why the extra protein is found so inconsistently in the field.

In nutrient solution cultures, the "nitrogen effect" can be found consistently, especially in solutions where the nitrate is taken into the plant immediately after triazine applications (BRUNETTI *et al.* 1971/72, GRAMLICH and DAVIS 1967, LORD *et al.* 1970, PULVER and TWEEDY 1968, RIES *et al.* 1963 and 1969). Measurements of nitrate reductase (NR) activity showed that enzyme activity was definitely increased (ARMSTRONG and ANDERSON 1971, ASLAM and HUFFACKER 1973, BUDOI and RIES 1968, CHMIEL and RIES 1967, DECLAIRE *et al.* 1974, RIES *et al.* 1967 a and b), SCHULKE 1970, TERRILLON and PAYNOT 1973, WU *et al.* 1971 c). Various effects of triazine herbicides on nitrate reductase have been observed in field tests, depending on the dosage applied and on the age of the seedlings. EASTIN (1966) and ZUMFT (1970) found that atrazine had no *in vitro* effect on nitrate reductase. Since nitrate as the enzyme's substrate induces its synthesis (CHEN and RIES 1968, SMITH and THOMPSON 1971), the nitrate uptake is more likely the reason for the observed increase in the nitrate reductase activity than a direct effect on the enzyme by the triazines. All these effects are gradually lost following discontinuation of the triazine treatment (WRAY *et al.* 1970).

This absorbed nitrogen is readily converted to organic nitrogen and found in the amino acid and protein fraction. An increase in the amino acid pool of the plants by triazines has been reported repeatedly (BARANOWSKI *et al.* 1967, GLABISZEWSKI *et al.* 1966 and 1967, GNANARETHINAM and CARLES 1966, GRAESER 1967 a, GRAESER and GZIK 1970, GZIK and GRAESER 1972 a, IDIATULINA 1963, PLOSZYNSKI 1972, PLOSZYNSKI and ZURAWSKI 1971, SWIETOCHOWSKI and MIKLASZEWSKI 1962, SWIETOCHOWSKI *et al.* 1966, WIEDMANN 1970). ZWEIG and ASHTON (1962) treated bean leaves with 10 ppm of atrazine in the light, and obtained no radioactivity in sucrose from $^{14}CO_2$ but aspartate, gluatamate, malate, serine, and glycine were labeled. The distribution pattern resembled the CO_2-fixation in the dark, but the aspartic and glutamic acid fractions were significantly labeled in comparison to the controls. In further experiments, ASHTON and URIBE (1962) compared atrazine-influenced metabolism of ^{14}C-sucrose and ^{14}C-serine by red kidney beans in light and in dark experiments. Also from these experiments, changes in the relative distribution of the ^{14}C-compounds are quite evident, especially in the amino acid fraction.

In all the reports on protein increases or decreases in plants, there are few references with detailed information. In green plants, the chloroplast proteins seem to be affected by detrimental concentrations of the herbicide, whereas the mitochondrial, microsomal, and cytoplasmic protein fractions remain intact. Working with *Avena* plants and herbicide concentrations detrimental to this plant (5, 2, and 1 ppm atrazine or simazine) SINGH (1965) and SINGH and WEST (1967) found a sharp reduction in the chloroplast protein fraction. Under experimental conditions where the herbicide treatment resulted in a greening effect and an increase in total proteins, it was found that the structural proteins of the chloroplast were particularly higher than in non-treated controls.

As discussed in the chapter "Seeds," the most interesting finding was that the seed protein of various plants can be increased by treatment with triazines. The protein spectrum of rye plants and pea seeds which were treated with simazine was investigated by RIES *et al.* (1967 a), and those of pea and sweet corn seeds by SINGH *et al.* (1972 c). Acrylamide gel electrophoresis of proteins extracted from the biological material revealed that the protein pattern was not altered by simazine application. They noted that more of the same types of proteins were apparently synthesized.

The influence of triazines on the synthesis of proteins was followed up by measuring the rate of incorporation of ^{14}C-leucine or ^{14}C-phenylalanine into plant proteins (JONES *et al.* 1972, MANN *et al.* 1965, MORELAND *et al.* 1969, PULVER and RIES 1973, TRUELOVE *et al.* 1973, WU *et al.* 1972 a). WU *et al.* (1972 a) measured higher rates of incorporation of ^{14}C-leucine into the water-soluble proteins of leaf discs from pea and sweet corn plants after treatment with simazine, propazine, terbutryne, and prometone. PULVER and RIES (1973) found that $10^{-8}M$ simazine in nutrient solution caused increased ^{14}C-leucine incorporation into the water-soluble proteins of oat plants prior to an increase in nitrate uptake. The increase in proteins by simazine was at the expense of the carbohydrates. Hydroxy-simazine had no effect in this experiment. In corn plants, where the nitrogen effect was visible by "greener" leaves and also accompanied by lower weight, the incorporation of ^{14}C-glycine into the structure proteins of the chloroplasts was found to be highly increased, the protein pattern had not changed as proved by an amino acid analysis (EBERT and VAN ASSCHE 1970). In these leaves the greening symptoms were found to be caused by larger and heavier chloroplasts with higher RNA synthesis which, of course, might explain the acceleration of the protein synthesis in these leaves.

The information available to explain all these effects of the triazine herbicides on protein synthesis as a whole and especially the increased nitrate uptake by the plant is still quite fragmentary and full of hypothesis. Recently, FEDTKE (1974) tried to give an explanation for similar effects of the herbicide methabenzthiazuron [*N*-(2-benzothiazolyl)-*N*,*N*'-dimethylurea] on crop plants. As many of his results are also obtained with triazines, his hypothesis might be relevant for photosynthesis-inhibiting herbicides including triazines. He found that herbicide-treated plants behave like shade-adapted plants. The chloroplasts are bigger in size, the total amount of chlorophyll is increased, but the ratio of chlorophyll a/b in decreased. The synthesis of amino acids and soluble proteins is increased as is the nitrate level and the nitrate reductase activity in these plants. The reduction of the contents of reducible sugars is thought to be a primary cause of the shade adaptation reactions in the plant metabolism. He also found a reduction in malate contents; however, a causal relationship between decreases in malate and increases in nitrate concentrations could not be confirmed. Recent experiments showed that shading of the plant or application of herbicides (among others atrazine at 0.1 ppm) stimulate nitrate uptake and that it is coupled with a reduction of $^{14}CO_2$-fixation (FEDTKE 1976).

This shade-adapting theory works when the herbicide concentration is high enough to reduce CO_2-fixation. The direct metabolic link between reduction of CO_2-fixation and growth stimulations remain an open question today. BUSH and RIES (1974) found that the elongation of the embryonic axis of red kidney beans by $10^{-8}M$ of atrazine was accompanied by an increase in protein synthesis. This effect was light-independent. The increase of protein synthesis, as measured by the incorporation of ^{14}C-leucine into the soluble protein fraction, began 1 to 2 hours after atrazine treatment. Stimulation of RNA synthesis, as measured by an increased incorporation of ^{14}C-ATP into the RNA fraction, began 3 to 4 hours after atrazine treatment. The metabolic reason for these increases in protein synthesis remains without answer at this time. The authors excluded from their results direct effects of the triazines on RNA metabolism as primary site of action during growth stimulations by triazines. Increased protein synthesis which is accompanied by an increase in total dry weight of the plant material is necessarily correlated with higher nutrient uptake.

Some observations have indicated that subtle membrane changes occur which could facilitate greater ion uptake. KRIHNING (1965) found suction pressure and water permeability were changed in the protoplast of epidermal cells from *Solanum tuberosum* petioles after treatment with simazine. Certainly the water balance of the plant can be affected by the triazines as discussed in chapter on Water Balance, and this alone may account for some of the observed changes in ion uptake. In some cases root growth is increased after applications of sublethal rates of triazines, and this would contribute to extra ion uptake.

Lastly, an alteration of the distribution pattern of the nitrogen in the plants by triazines may aid in explaining their effects on yield and proteins. In rye seedlings, the shoot-root ratio is unaffected by atrazine; however, the nitrogen content rises in the shoot and concurrently falls in the roots (DILL and CARTER 1971 and 1972, DILL 1973). Also, other scientists confirm that nitrogen and protein increases occur predominently in the aerial parts of the plants (PLOSZYNSKI 1973). All these factors may contribute to the buildup of a larger biomass in the plant.

b) Nutrients and chemicals influencing the phytotoxicity of triazines

Increasing rates of NPK-fertilizers were shown to enhance the phytotoxicity of triazines (BOIKO and PETELKO 1966, VESELOVSKII and MAN'KO 1969), and the concentration and form of the nitrogen had an influence on their phytotoxicity. High nitrogen levels, in general, favored their herbicidal activity (DOLL and MEGGITT 1969, FIGUEROLA 1970), especially when nitrate was applied as fertilizer (DOLL 1969). Depletion of carbohydrate reserves in the rhizomes of quackgrass was substantially greater on plots to which nitrogen had been applied (SCHIRMAN and BUCHHOLTZ 1966). Less herbicidal injury was obtained in soybeans which were fertilized with ammonium sulfate than in those fertilized with calcium nitrate (MCREYNOLDS and TWEEDY 1970). Corn, rye, and soybean plants took up only half the amount of ^{14}C-simazine from a nutrient solution containing ammonium nitrogen as compared with one containing nitrate nitrogen (MCREYNOLDS and TWEEDY 1969).

In addition to these interactions between triazines and nitrogen uptake, a phosphorus-triazine interaction has been discussed in many reports (ADAMS and OTTO 1964, DHILLON *et al.* 1967, DOLL *et al.* 1970, JANISEVSKIJ and GRICEVIC 1967, RUDGERS *et al.* 1970, SUN 1970, SUN and ADAMS 1971, UPCHURCH *et al.* 1963, STOLP and PENNER 1973). High levels of phosphorus in soils lead to increased susceptibility of plants to triazines (ADAMS 1965, ADAMS and ESPINOZA 1969, ADAMS and SUN 1971, DOLL *et al.* 1970, SUN 1970), although no extra herbicide is taken up. Since high phosphorus levels are detrimental to chloroplast development this interaction can be expected. SUN and ADAMS (1971) showed that the subcellular effects on soybean chloroplasts by manganese deficiency, by excess of phosphorus, or by atrazine were the same. This suggested that both phosphorus and atrazine were interfering with the function of manganese in membrane structure and photosynthesis.

PENNER (1970) investigated the effects of triazine on phytase and found it somewhat inhibitory even in dark grown barley seedlings. Such an inhibition could decrease the internal inorganic phosphate level since phytin levels would remain high in germinating seeds.

A similar case exists for magnesium-triazine interaction. Magnesium-deficient plants are more susceptible to triazine injury and added magnesium will partly overcome their toxic effects (BRENCHLEY 1969, BRENCHLEY and APPLEBY 1967 and 1971). The role of magnesium in chlorophyll can explain these interactions; however, at this time, no molecular mechanisms can be proposed for these interrelationships.

Besides the mineral nutrition, other chemicals and pesticides may contribute also to the triazine toxicity in plants. A synergism between the triazines and other types of herbicides as amitrole, 2,4-D, and others, has been reported repeatedly (AKOBUNDU *et al.* 1972; BEREZOVSKII *et al.* 1965; DIEM 1971; DIEM and DAVIS 1969, 1970, and 1974; DOROZKINA *et al.* 1973; DUKE *et al.* 1972; GRUZDEV and KALININ 1967; LYNCH *et al.* 1970; HAGIMOTO and YOSHIKAWA 1972). Soybean plants were more injured by ametryn when 2,4-D was present in low concentrations (DIEM and DAVIS 1970 and 1971). Auxins seem to exert a synergistic or antagonistic effect on the triazine action according to their concentrations and the plant species under investigation (BEREZOVSKII *et al.* 1965). DOROZKINA *et al.* (1973) showed that the antagonistic action of 2,4-D on atrazine toxicity in beans is due to a change in the distribution pattern of the assimilates by this chemical. Likewise, plant hormones such as indole-3-acetic acid and gibberellic acid had an influence on the uptake of propazine into leaves of *Camilla sinensis* plants (BISWAS and HEMPHILL 1965).

Increases in phytotoxicity of simazine to *Agrostis alba* were obtained with additional applications of auxins and gibberellins (DEVLIN 1967). Several organophosphate insecticides were shown to alter the phytotoxicity of triazines (CORBIN *et al.* 1970, PENNER 1974). The insecticide had an influence on the absorption and distribution of ^{14}C-atrazine in the plant.

The search for chemicals which reduce the triazine toxicity in plants has led to many investigations. Activated charcoal is known to adsorb chemicals as the triazines and other herbicides resulting in reduced uptake of the chemical by the plant and protecting the plant in this way (GAST 1956 and 1962, ALLOTT 1968, DVORAK and STANEK 1968, LINSCOTT and HAGIN 1967, LONG and SCRANTON 1969; see also GUNTHER and GUNTHER 1970). KRATKY and WARREN (1969) found that carbon afforded protection against atrazine in *Chlorella pyrenoidosa*. Also a growth regulator has been investigated in this respect. KIRKLAND (1973) found that CCC gave some protection against atrazine injury in wheat plants. He found that the inhibition of net photosynthesis by atrazine was slowed down in CCC-treated plants as compared to controls without CCC. Possibly the stabilizing effect of the CCC on chlorophylls can account for this safening effect.

c) Susceptibility of plants against phytopathogens

Several papers also report on interrelationship between the effects of the triazines on plant physiology and the susceptibility of the plant against attack by parasites. This chapter of secondary triazine effects seems to be rather neglected although investigations of this kind could be of economic importance in integrated pest control. HEITEFUSS (1970 and 1972), VAN DER ZWEEP (1970), and MOORE and THURSTON (1970) have summarized these observations of pesticides, including the triazine herbicides. HEITEFUSS and BODENDOERFER (1970) found that the attack of simazine-treated wheat plants by the fungus *Cercosporella herpotrichoides* was reduced considerably in comparison with untreated controls. An explanation of this finding is very complex. Decrease in the carbohydrates and increases in the nitrogen (protein and amino acid) fraction in the triazine-treated plants may, of course, contribute to favorable or unfavorable nutrient conditions for the development of the fungus. However, the increased buildup of fungi-toxic compounds like the DMBO glucoside (4-hydroxy-7-methoxy-1,4-benzoxazine-3-one-2-glucoside), which is enzymatically converted in the plant to the fungitoxic aglucone (2,4-dihydroxy-7-methoxy-1,4-benzoxazine-3-one) has to be taken into consideration, also. A change in the hormonal and development status of the plants may contribute to their resistance, too (HEITEFUSS 1970 and 1972, HEITEFUSS and BODENDOERFER 1970, BRANDES and HEITEFUSS 1971 a and b). The hormonal status of conifer plants seems to play a role during infestation by *Lophodermium pinastri.* Needle retention was positively influenced by simazine in the more vigorously growing conifer plants. These plants were more resistant against needle fall which is caused by the fungus (JANCARIK 1965).

Several herbicides are known to affect healthy and virus-infected plants to different degrees (HEATHCOTE 1970). Simazine intensified the symptoms (chlorosis, stunting, etc.) caused by the sugarcane mosaic virus without affecting healthy sugarcane (ADSUAR 1961). In tobacco, simazine was lethal only to virus-infected plants (ULRYCHOVA and BLATTNY 1961). The authors proposed using a simazine spray to detect tobacco mosaic virus in diseased plants. SCHUSTER (1971) has tried to develop a technique for virus diagnosis in potatoes using simazine or atrazine; however, he found the technique to be hardly suitable for practical purposes. The widespread occurrence of the maize dwarf virus in several corn production areas and possible interactions with triazines altering the resistance against the virus was the subject of a study by MILLIKAN and MANN (1969). They found no deleterious virus-simazine interactions of maize dwarf virus on a virus-susceptible and a virus-tolerant variety of *Zea mays*. In virus-inoculated plants, inhibitory effects of triazines with respect to total plant growth and $^{14}CO_2$ fixation appeared to be additive with the virus effects (Foy *et al.* 1969).

These few examples show that it is worthwhile studying metabolic interrelationships in plants when combination products are used, or when techniques for integrated pest control are considered.

IV. Nucleic acids

Effects of triazine herbicides on plant growth and protein synthesis can also be reflected in RNA-metabolism. Some results from experiments with higher plants have indicated that this is the case for both stimulations and inhibitions of RNA-synthesis (Graeser 1970, Kljajic 1962, Singh and West 1967, Solecka *et al.* 1969). Relationships were found between their nucleotide (ATP, ADP, AMP) content and plant growth. Concentrations of simazine which stimulated plant growth also gave a higher nucleotide content (Graeser 1967 b). From a study on the effects of a series of different herbicides on ATP levels in plant tissue, it was concluded that any herbicide which affects energy-linked processes such as ATP production also will affect indirectly RNA and protein synthesis (Gruenhagen and Moreland 1971). This seems to be a general rule and is certainly true for the triazines which were shown to influence systems providing inorganic nutrients such as phosphate or nitrate to the plant cell thus changing the basis for metabolic processes.

The induction of nitrate reductase activity by additional nitrate or supplementary ATP available in the cell could be a first step in changes which lead to effects on RNA-synthesis. Enzymes like ATPase and nitrate reductase were enhanced after spraying bean plants with growth stimulatory low amounts of several triazines (Singh and Salunkhe 1970). The hypothesis to explain the effects of triazines on nucleic acid synthesis primarily as a nutrient effect has some support, because it was shown that increases in oat grain protein as well as qualitative changes in the composition of these proteins were the same from simazine treatments as from additional nitrogen fertilizer treatment without simazine (Steenbjerg *et al.* 1972).

Furthermore, it is well documented that the reported drastic stimulations or inhibitions of RNA and protein synthesis occur mainly in intact plants with access to a nutrient pool, such as the soil, or to storage reserves from the seeds. They do not occur in explants. The effect of atrazine on the synthesis of RNA and proteins was investigated along with a series of other herbicides in excised tissue assays (Hoogstraten and Ashton 1972, Mann *et al.* 1965, Moreland *et al.* 1969). Incorporation of ^{14}C-leucine into the proteins of plant material as a measure of the *de novo* synthesis of proteins was not affected by atrazine (Mann *et al.* 1965, Moreland *et al.* 1969), nor was the induction or activity of gibberellin-induced-α-amylase (MORELAND *et al.* 1969). This last is taken as a measure for effects on RNA synthesis mechanisms through which gibberellin must act to induce α-amylase production. Incorporation of ^{14}C-ATP as a measure of RNA-synthesis was inhibited and that of ^{14}C-orotic acid as a precursor of pyrimidines was stimulated by atrazine at concentrations of $2.10^{-4}M$ (MORELAND *et al.* 1969). Ribosomes and enzymes from cotyledons of etiolated pea seedlings were used to measure incorporation of ^{14}C-phenylalanine into proteins, but atrazine had only a minor effect (HOOGSTRATEN and ASHTON 1972).

Attempts have been made to explain the effects of triazines on nucleic acids at the level of nucleic acid synthesis. Because of structure similarities, it was natural for some researchers to postulate that the triazines might possibly substitute for the pyrimidine bases thymine and uracil. Considerable attention has been paid to this hypothesis during the last few years and several studies have been undertaken to determine possible interactions of triazines with nucleic acids in this respect.

Auxotrophic bacteria were preferable test organisms for elucidating the mode of action of triazines on nucleic acids since secondary effects were eliminated which could arise from the more complex situation in higher plants, e.g., by interfering with ATP-production from energy-transferring processes. The effects of 12 triazines were investigated on the growth of *Bacillus subtilis* A-26, an auxotroph which requires uracil for growth. The study was limited to growth studies as an index of *s*-triazines substitution for uracil. The data showed no evidence that the *s*-triazines could serve as precursors of uracil synthesis or serve as biological analogues to uracil in this organism (ERCEGOVICH and HERENDEEN 1967). The growth rate of a supposedly mutant strain of *Escherichia coli 15 arg⁻t⁻u⁻* which requires arginine, thymine, and uracil for growth was reported as improved by several triazines (ERCEGOVICH *et al.* 1966), but later the bacteria were revealed to be *Mima polymorpha* strain with a preference to grow on ethanol. Therefore, the growth response which was observed was due to the ethanol used to dissolve the triazines and not to the triazines themselves (ERCEGOVICH 1968). The *E. coli* $arg^-t^-u^-$ strain was later used for studying possible incorporation of ^{14}C-prometryn and ^{14}C-cyanuric acid into the nucleic acid fractions. Low amounts of radioactivity were detected in the nutrient deficient but not in the complete medium. However, a positive control was not included in this test (TEMPERLI *et al.* 1966), and the result could not be repeated by other investigators using the same test system (MUECKE *et al.* 1969). Ring-labeled atrazine and prometryn were not incorporated as base analogues into DNA or RNA, but 2-^{14}C-5-bromouracil as a positive control was highly effective. There was no incorporation of C-prometryn into the total nucleic acid fraction of *E. coli* S-26 during the growing phase nor did the incorporation of $^{32}PO_4$ differ from the controls (SANSING and CHO 1970).

Within a series of herbicides, several triazines of the chloro-, methylthio- and methoxy-type were evaluated for possible mutagenic properties in a microbial system. The test system consisted of eight histidine-requiring mutants of *Salmonella typhimurium*, bacteriophage T_4, and 2 rII mutants of bacteriophage T_4. This system is reported to be sensitive in detecting genetic damage of the point mutation type, to single genophores or to the DNA molecule. The triazines like most of the other herbicides evaluated were ineffective in this test (ANDERSEN *et al.* 1972). Several of these histidine-requiring strains were used to check the increase of the number of spontaneous reversions. Among several herbicidal triazines, triazine metabolites, such as hydroxysimazine, ammeline, or melamine were also tested. They showed no activity in this particular test. Diethyl sulfate as a positive control gave good results (SEILER 1973).

The influence of atrazine on nucleic acid and protein synthesis of *Neurospora crassa* was also investigated (SCHROEDER *et al.* 1970). Atrazine had an insignificant influence on RNA or DNA synthesis in this organism, although protein synthesis was enhanced by 40%. The induction of respiration-deficient yeast mutants indicates effects of chemicals on DNA, but atrazine had no effect in this test either (SCHUBERT 1969).

Not only the incorporation of the chemical in question into the nucleic acids must be considered, but physical effects, such as intercalations with the bases in DNA, or allosteric binding to proteins, which are important in nucleotide or nucleic acid metabolism, have to be considered as possibly causing biological effects. Several types of possible physical interactions of triazines with nucleic acids have been investigated (SANSING and CHO 1970). The possibility of nonenzymatic binding (possible physical interactions) of ^{14}C-prometryn to a single or double stranded phage DNA (T_5-phage or M-13 phage DNA) was investigated by using equilibrium dialysis, denaturation (heating and reannealing) of DNA, and MAK chromatography. In no case was there evidence of an attachment of the triazines to nucleic acids. Furthermore, viral production should be affected if the infected bacteria were growing in a medium containing a chemical with intercalating properties. Phage yields as determined by plaque assays from bacteria grown in atrazine, simazine, or prometryn solutions showed no significant difference from controls. However, acridine orange, an intercalating dye, completely inhibited viral production in this test. Possible effect of triazines on nucleic acid transformation were investigated as well. Again, the triazines had no effect when added with the transforming DNA in a system of *Bacillus subtilis* auxotroph for tryptophan being transformed a prototrophic organism (SANSING and CHO 1970).

Also, from studies with higher plants, no incorporation of the ring moiety from ^{14}C-atrazine was found in the nucleic acid fractions of resistant and susceptible selections of corn (*Zea mays* L., GT-112-R and GT-112-S) (SCHULTZ and TWEEDY 1968). Association of ^{14}C-simazine with nucleic acids has been reported from an experiment with corn seedlings (GRAESER 1969 a). In this experiment, root application of ring-labeled simazine to etiolated corn seedlings was followed by nucleic acid and nucleotide fractionation studies. Radioactivity from simazine was found in all nucleic acid fractions without showing specificity. All nucleotide fractions contained label as well, but a specific nucleotide of simazine or simazine metabolite was not identified (GRAESER 1970). Therefore, discussions as to what was the nature of the labeled material remain open.

If triazines are operating at the level of RNA synthesis, antagonists of nucleic acid or protein synthesis, such as 8-AA (8-aza-adenine), 5-BU (5-bromouracil), CA (chloramphenicol), and 5-FU (5-fluorouracil) should nullify growth stimulations by triazines or they should increase their inhibiting action. This was shown to occur in etiolated corn seedlings (GRAESER 1969 b). A biochemical study was undertaken to examine the changes of RNA content in plants by simazine and 5-FU (GRAESER and TILLICH 1970). Nucleic acids from the primary roots were characterized. Incubation of the seedling with $5 \times 10^{-4}M$ simazine previously shown to influence root growth resulted in a strong depression of the amount of soluble RNA and DNA-RNA fraction but the ribosomal RNA fraction was significantly higher. The incorporation of ^{32}P into the nucleic acids was increased in all fractions. In this test system, 5-FU inhibited soluble RNA and DNA-RNA synthesis. Simazine reversed this inhibition and produced a stimulation of RNA synthesis. From these studies, the authors concluded that the specific site in metabolism through which the triazines influence *de novo* synthesis of RNA and/or proteins is not the same as that of the well-known antagonists of nucleic acid and protein synthesis. The exact site involved remains unknown, because conclusions from these tests with whole plants are complicated by the effects of triazines on other plant systems, such as ion uptake and plant hormones, which lead to side effects on RNA-synthesis. Induction of nitrate reductase activity is preceded by *de novo* RNA-synthesis. Auxins are known to influence RNA contents, especially in the ribosomal nucleic acid fraction. All these aspects should be considered when interpretations of the research on this topic are made.

An interesting study was undertaken by PENNER and EARLY (1972) in which they isolated chromatin from etiolated soybean seedlings which contained DNA template and sufficient RNA polymerase to support acid-precitable RNA synthesis in the presence of all four nucleotide bases and manganese. Atrazine applied to the seedlings 6 hours before harvest enhanced RNA synthesis. If atrazine was added to the reaction mixture after the extraction procedure, RNA synthesis was enhanced also. The authors came to the conclusion that the action of atrazine in this system was a de-repression of transcription by increasing the template availability.

Several cytological studies have been undertaken to check possible effects of triazines on chromosomes. In *Vicia faba* stipular cells and in *Tradescantia* sp. hair cells which were treated with low concentrations of atrazine, mitosis proceeded and finished normally from early prophase

(SAWAMURA 1965). In root tips of *Vicia faba* and *Hordeum vulgare* as well as in pollen mother cells of *Sorghum vulgare*, cell division was normal under the influence of the triazines. The rate of cell division was not changed, and there was no deviation in the structure of the chromosomes (MUELLER *et al.* 1972). No mutagenic effects occurred following two generations of *Arabidopsis thaliana* (LEUTHOLD 1971). In barley plants, simazine and atrazine did not increase the proportion of sterile spikes nor induce any chlorophyll mutants (STROEV 1968 a). Cytogenetic activity in the roots of spring barley was not altered by simazine treatment (STROEV 1968 b). Atrazine seemed to interfere with meiotic stability in *Sorghum vulgare* to a low extent; however, no apparent relationship existed between these effects and yield responses following atrazine application (LIANG *et al.* 1967, LIANG *et al.* 1969, LIANG and LIANG 1972). Slight activity of triazines in tests with *Vicia faba* and *Hordeum* root tips and *Hordeum* pollen mother cells (WUU and GRANT 1966 and 1967 a and b) was not repeated in complementary tests (MUELLER *et al.* 1972).

In this respect the paper by KIERMAYER in the book edited by L. J. Audus "The Physiology and Biochemistry of Herbicides" (1964) gives a misleading interpretation of the papers by WAKONIG and ARNASON (1958 and 1959) with the statement that the triazine herbicides would cause chromosome breakage like the alkylating agent TEM [2,4,6-tri-(ethylene-imino)-1,3,5-triazine], a compound which has been used in cancer chemotherapy on account of its alkylating properties. The triazine herbicides are not alkylating agents like TEM. The possibility of mutagenic effects from several triazine derivatives has been investigated. The triazine ring itself as well as melamine (2,4,6-triamino-1,3,5-triazine) is genetically ineffective. In comparison, trimethylol-melamine [2,4,6-tri(oxymethylamino-1,3,5-triazine] and TEM have a strong alkylating action of the three methylol and ethylene substituents, respectively (ROEHRBORN 1962).

When all the results from experiments of effects of triazines on nucleic acids in plants are considered, there is no solid evidence for any type of direct interference from the triazine herbicides. This was shown in microorganisms, isolated explants, and higher plants. Stimulations of nucleic acid synthesis have been reported mostly from intact higher plants which were growing either in a nutrient medium or from young seedlings with nutrient reserves in the seed, and direct measurements of nucleic acid metabolism could not be made in such experiments. The hypothesis that a triazine may compete with a base analogue in the buildup of nucleotides could not be confirmed in any case, though intensive work has been done in this respect over several years. However, another hypothesis seems more appropriate to explain the apparent triazine action on nucleic acid and protein synthesis in higher plants: the effects are indirect. As it is well established that the triazines have an influence on systems which provide additional nutrients through the roots or from storage reserved for the seedling, these additional nutrients could be thought to change the metabolic basis and energy requirements even before the RNA-level and lead to visible effects in higher plants. Also the aspect of influencing RNA-metabolism through effects on plant hormone content should be considered as an explanation for the growth reactions of plants to triazines.

V. Morphology and plant growth

a) Injury symptoms

Any herbicide will result in changes of plant growth behavior at the microscopic level first and then visually at the macroscopic level. In the case of the triazines, where the inhibition of photosynthesis is the main metabolic interaction which leads to plant destruction, chlorosis is the first appearing principal type of injury symptom (Lynd *et al.* 1967, Wong and Romanowski 1968). Veinal chlorosis of the palmate type appears two to three weeks after soil application, but following foliar applications it begins after several days at the leaf blade edges and spreads over the whole leaf surface (Gast and Knuesli 1956, Gast 1959 b, Lynd *et al.* 1967).

The typical triazine chlorosis appears with CO_2-inhibition effects (Sasaki and Kozlowski 1967 b), but at the cellular level structural changes precede the visible symptoms. Chlorosis is usually preceded by the appearance of water-soaked lesions beginning near the edges of the lamina or around the veins and spreading over the whole leaf surface until it has a water-soaked appearance. When structural changes in plant tissue preceding chlorosis were investigated microscopically, ultrastructural changes occurred only in chloroplasts, whereas mitochondria appeared to be intact (Hill *et al.* 1968). However, in algae atrazine did not cause any observable abnormalities in cell organelles with little, if any, effect on the structure of the chloroplast, nucleus, mitochondria, or endoplasmatic reticulum (Ashton and Bisalputra 1964, Ashton *et al.* 1966). Fong and Norris (1972) investigated the intracellular localization of ^{14}C atrazine and ^{3}H-atrazine in corn, bean, and oat plants. They found by sucrose density gradient centrifugation an accumulation of the radioactivity in the chloroplast fractions from the sensitive oat and bean plants, but not from the resistant corn plants. This preferential accumulation of atrazine did not occur in the other organelles. The distinct localization of radioactivity from ^{14}C-atrazine in the chloroplasts was confirmed by microradioautography.

The morphological breakdown of chloroplasts, which occurs after triazine treatment in light, was followed by electron microscopy (Ashton *et al.* 1963 b). As early as 30 hours after application of the herbicide, the initial change in the morphological shape of chloroplast took place, turning from discoid to spherical shape, and after 96 hours all stages of chloroplast destruction occurred. After the initial disappearance of the starch from the lamellar system, the frets were destroyed and the grana arrangement became disorganized. The compartments of the grana became swollen and disintegrated (ASHTON *et al.* 1963 b). Thylakoids seem to be less sensitive; the herbicidal effect on the thylakoid units apparently is limited to membrane swelling (CAMPBELL and EVANS 1971). The age of plants or the varying plant physiological development stages change the susceptibility of plants towards triazines, but still the particular characteristic chloroplast changes were not altered. For example, barnyardgrass varies in its susceptibility to atrazine at different growth stages (HILL *et al.* 1968), but the chloroplast changes caused by triazines were the same in all growth stages.

Structural changes in plant tissue which accompanied chloroplast breakdown have been investigated also. The vascular tissue of the stems of bean plants which were exposed for 4 days to nutrient solution containing 10 ppm of atrazine showed a strong cessation of growth in the cambium zone. Sieve elements of the treated plants had thinner walls. Recently formed primary and secondary xylem had thinner walls, too, partly collapsed. The number of the cell layers within the mesophyll remained the same, but the interior airspace-system was not developed extensively. All these effects were observed only in treated plants which were exposed to the light, but not in treated plants which were kept under dark conditions. There seems to be a definite requirement for both the presence of a triazine and light before a progressive disintegration of the chloroplast takes place as noted during these experiments (ASHTON *et al.* 1963 a).

Chlorosis is the only visible symptom of the phytotoxic action of the triazines in plants. No abnormalities in the anatomical structure or malformations in plant organ development and plant morphology have been observed (GERMANO 1965). In shoot apices leaf development was investigated at different stages, which included early differentiation, enlargement of palisade parenchyma and early maturation. During a 96-hour period, no differences occurred between untreated control plants grown under light and atrazine-treated plants kept under dark conditions except an increase in vacuolation which also occurred in untreated plants in the dark (ASHTON *et al.* 1963 a). One author reported anatomical responses of pine (*Pinus resinosa*) needles to treatments with atrazine, simazine, prometryn, and propazine (KOZLOWSKI and CLAUSEN 1966). Herbicide-treated plants were described as developing distorted needles which were triangular in cross sections, whereas cross-sections of needles from control plants were approximately semicircular. However, from these findings, a misleading interpretation was drawn. Since the authors had illustrated and described three different kinds of foliar appendages of red pine during one experiment (KRUGMAN and CRITCHFIELD 1968), it is difficult to derive an influence of the triazines on leaf anatomy from this experiment. Following 3 generations of *Arabidopsis thaliana*, chronically treated with atrazine, no effect on morphology or induction of genetic aberrations could be observed (Leuthold 1971), so it seems that permanent changes in plant anatomy as a result of triazine treatment do not occur.

b) Growth stimulations and inhibitions

Since the early field work with triazines, it had been observed repeatedly that triazine herbicides at sublethal concentrations could exert an influence on the growth of plants which is totally apart from any benefit due to weed control. Stimulations of growth, as well as inhibitions, were observed, and a voluminous amount of literature appeared on the subject. These stimulations of plant growth did not only affect shoot length, but the leaf blade surface, the stem thickness, and root growth were all increased. Stimulations of shoot growth after triazine treatment were observed in corn (Bartley 1957, Freney 1965, Lorenzoni 1962, Roth 1959, De Vries 1963), and were shown to be dependent on temperature (Chiesura and Lorenzoni 1964). Furthermore, shoot growth enhancement also occurred in wheat, barley, and rye (Lorenzoni 1962), in sorghum (Mannikar *et al.* 1971), vines (Gast 1960, Gast and Grob 1964), in orchards, mainly apple and peach trees (Gast and Grob 1964, Karnatz 1964, Karnatz 1965, Mohs 1961, Ries *et al.* 1963), cotton (Idiatulina 1963), conifers and woody plants (Heid 1962, Jancarik 1965, Mueller 1966), citrus (Goren and Monselise 1966), and 25 different ornamental species (Petojan 1966). The leaf blade surface was increased in fruit trees (Gast and Grob 1964, Karnatz 1964), winter wheat (Hafner 1963 and 1964), corn (Graeser 1966), peas (Schulke 1970), and summer linds (Heid 1962). Trunk diameter was improved in fruit trees (Karnatz 1964, Larsen and Ries 1960, Ries *et al.* 1963). Root growth was positively influenced in pine (Lund-Hoeie 1972), cress (Ebert and Van Assche 1969), corn (Graeser 1967 b), oats (Wiedman and Appleby 1972), and fruit trees (Dvorak 1968). Spring wheat, millet and corn were suppressed in growth by simazine and atrazine, but the chemicals stimulated initial root size (Mashtakov and Prokhorcik 1962 c). Rooting of cuttings was improved in beans (Davis and Copping 1970, Iliev 1973). At the higher subherbicidal triazine concentration inhibitions of plant growth have been observed (Dvorak 1968, Goren and Monselise 1966, Iliev 1973, Mueller 1966, Swietochowski *et al.* 1962, De Vries 1963). These growth responses of plants to triazines, as seen from the reported literature, include many plant species, triazine-sensitive as well as triazine-resistant species and annuals as well as perennials. However, a specific concentration is required to produce these effects, and this varies greatly from one species to another, even according to the specific development stage of the species under investigation.

Growth reactions of plants treated with triazines were also found when the emerging force of seedlings was measured. It was significantly higher in soybean plants which grew more vigorously after a treatment with atrazine (BURNSIDE 1971). Plants undergo a specific rhythm with respect to stimulations of inhibitions. In a bioassay, it was shown that oat plants which were treated with 0, 0.5, 1, 1.5, 2, 2.5, and 3 kg/ha simazine were stimulated in growth during the first days. A neutral phase followed, and finally inhibition of growth occurred. At the higher simazine concentrations, similar phases of growth reactions occurred as at the lower concentrations but in a shorter time period (SWIETOCHOWSKI and MIKLASZEWSKI 1962, SWIETOCHOWSKI *et al.* 1962, SWIETOCHOWSKI *et al.* 1964).

Changes in the plant growth behavior have been studied repeatedly at the cellular level (ASHTON *et al.* 1963 a and b, FOY and BISALPUTRA 1964, RUBIN and GRIZAJENKO 1964). The microscopic picture of stem sections of corn after triazine treatment revealed that the number of sclerenchyma fibres in the xylem and the diameter of the supporting tissue in the cortex were increased (RUBIN and GRIZAJENKO 1964). In cotton leaves, anatomical responses to prometryn were noticed where subherbicidal concentrations caused the treated leaves to become thicker and darker green in appearance. This was due to the development of one to several extra tiers of spongy mesophyll (FOY and BISALPUTRA 1964).

Contrary to higher plants, the growth of algae such as *Chlorella* is not or is only slightly influenced by simazine and/or atrazine (HELLING *et al.* 1971, LEOPPKY and TWEEDY 1969, THOMAS *et al.* 1973, VANCE and SMITH 1962. Since glucose is reverting the inhibiting effect of atrazine on this algae (ASHTON *et al.* 1966), it was suggested that inhibitions of growth which were observed in *Chlorella* after treatment with atrazine (ASHTON *et al.* 1966, GRAMLICH and FRANS 1964, HELLING *et al.* 1971, KRATKY and WARREN 1971, KRUGLOV 1970, WEIDNER and LAVY 1973), and simazine (KRATKY and WARREN 1971) may be explained mainly by inhibition of the Hill reaction.

Growth responses of plants to triazines occur not only in green but also in etiolated plant material (JORDON *et al.*, 1966, GRAESER 1966 and 1967 b, GZIK and GRAESER 1972 b, ILIEV 1973, BUSH and RIES 1974). No direct correlation can be found between the triazine action on photosynthesis and growth response. The effects of the triazines on plant hormones and plant growth are based on their effects on protein and RNA-metabolism (see previous two chapters).

c) Interactions with plant hormones

Several observations lead to the conclusion that plant hormone metabolism may be influenced by the triazine herbicides. In order to determine the type of plant hormone affected by the triazine herbicides, the influence of atrazine on auxins, cytokinins, and gibberellins has been checked. The influence of atrazine on the gibberellic acid metabolism was invesigated by using the barley endosperm biotest, and no effect was seen (EBERT and VAN ASSCHE 1970, MORELAND *et al.* 1969). Effects of several triazines on opening of hypocotyl hook of cotton make one think of ethylene interactions in plant metabolism (DAVIS and PILLAI 1972, PILLAI and DAVIS 1973 a). Auxins as well as cytokinin metabolism seem to be involved in the triazine action.

Auxin (IAA) metabolism was shown to be influenced by triazines. Wheat coleoptiles and cress roots reacted positively to low levels of atrazine (EBERT and VAN ASSCHE 1969).

Coleoptile growth of etiolated corn seedlings was influenced by simazine, atrazine, and propazine (GRAESER 1966). For a biochemical test to detect changes in auxin metabolism, an IAA oxidase system has been used which measures the rate of decarboxylation of radio-labeled IAA in etiolated oat coleoptiles. In wheat and oat coleoptiles, a strong influence of atrazine could be demonstrated, but not in an *in vitro* system (EBERT and VAN ASSCHE 1969). Concentrations of $10^{-6}M$ atrazine stimulated decarboxylation whereas concentrations lower than $10^{-12}M$ inhibited it. Peroxidase also was affected by simazine in *Brassica* seedlings (BUSCHBECK *et al.* 1971).

Root growth, which is considered to be controlled mainly by auxins, was affected which supports the idea that the triazines change the auxin level in the plant. Atrazine and propazine inhibited the growth of root tips of *Lens culinaris,* increasing the partial growth inhibition caused by IAA (PILET and GASCHEN 1962). The same is true for $10^{-4}M$ concentrations of atrazine and ametryn which inhibit the growth of excised tomato roots (COPPING 1971, COPPING *et al.* 1972). The rooting of cuttings from simazine-treated plants which were dipped into a 100 ppm of NAA-solution to facilitate rooting was retarded in comparison to untreated controls (MCGUIRE and PEARSON 1972). Root initiation on mungbean hypocotyls was increased by $10^{-6}M$ atrazine or ametryn in the absence of exogenous applied IAA but in the presence of 10 ppm of IAA, root initiation was inhibited. Root initiation in soybean callus was observed by $10^{-14}M$ ametryn plus 10 μg/ml kinetin (COPPING *et al.* 1972). If the assumption is right that root initiation is induced at a high auxin-to-cytokinin ratio, the effect of the triazines could be due to increased auxin or decreased cytokinin levels in the callus, and we have already observed that IAA decarboxylation is inhibited in plant tissue by low concentrations of the triazines (EBERT and VAN ASSCHE 1969). However, there may be other explanations to consider as well.

An early positive influence of triazines on root growth would explain growth stimulations of plants, since the enlarged root-surface would increase nutrient uptake. Stimulations of root growth by subherbicidal levels of triazines were reported for cress (EBERT and VAN ASSCHE 1969, GRAESER 1967 b), oats (WIEDMAN 1970, WIEDMAN and APPLEBY 1972), corn (GRAESER 1966), and conifers (LUND-HOEIE 1972). Root growth was shown to be stimulated more than shoot growth. In this respect, an interesting study was performed with *Picea abies* seedlings, which demonstrated regulator-like effects of triazines. Following an application of simazine in the stage of dormancy, an increase of root growth and an inhibition of shoot growth was observed during the first year. In the second year, shoot growth and needle length were enhanced and needle color was improved significantly. Maximum stimulation occurred during the third and fourth year after simazine application (LUND-HOEIE 1972). Simazine was shown to influence the basipetal transport of ^{14}C-assimilates in *Picea abies* (LUND-HOEIE 1969 a and b). The different distribution pattern of assimilates and possibly also auxin could induce changes in growth reactions of the simazine-treated plants especially the year of treatment. Thereafter, plant growth may be stimulated as a result of a more developed root system.

Possible effects of triazines on cytokinin metabolism might be expected, since a "greening-effect" can be observed sometimes after triazine treatment (BARTLEY 1957, GAST 1960, ILIEV 1973, KARNATZ 1964, GOREN and MONSELISE 1966, NEWTON 1966, NEWTON and WOODARD 1965). A cytokinin-like activity from atrazine was implicated in causing a delay in senescence of corn leaves (FOY and HIRANPRADIT 1970, HIRANPRADIT and FOY 1973). Green color in Christmas tree plantations (*Pseudotsuga taxifolia BR*) was improved by atrazine treatment (NEWTON 1966, NEWTON and WOODARD 1965). Needle retention was positively influenced indirectly by simazine in the more vigorously growing conifer plants (JANCARIK 1965). This "greening-effect" from triazines is not caused through interactions of the herbicide with the microflora of the soil or the plant rhizosphere, since it was demonstrated on plants which were growing on completely sterile medium (EBERT 1968). In an attempt to explain the "greening-effect," chloroplast number, diameter, weight, protein, and chlorophyll were determined. The chloroplasts from atrazine-treated corn plants which showed the "greening-effect" had greater diameters, were heavier, and contained more structure proteins (EBERT and VAN ASSCHE 1970). Also, in single-cell algae under steady light conditions, atrazine increased the chlorophyll content, as did DCMU (BOEGER 1974).

Comparisons have been made between the action of atrazine and cytokinins on chlorophyll retention and chlorophyll synthesis in leaves. Chlorophyll retention studies from senescence biotests with leaf discs of *Rumex acetosa, Taraxacum officinale, Avenasativa, Sinapis alba,* and *Zea mays* gave no effect from atrazine under light conditions when compared to kinetin (EBERT and VAN ASSCHE 1970, COPPING and DAVIS 1972). However, corn (*Zea mays*), mustard (*Sinapis alba*), or radish (*Ralphanus sat.*) leaf discs which were incubated in the dark with either triazines or kinetin showed higher chlorophyll a and b contents than water-controls. Combinations of the two compounds decreased these chlorophylls (COPPING 1971, COPPING and DAVIS 1972, ILIEV 1973).

These studies showed significant interactions between atrazine and kinetin on the chlorophyll a and b contents of corn leaf discs. Those interactions between cytokinins and atrazine on chlorophyll formation are confirmed in another study (PENNER and WILEY 1972). In cotyledons excised from squash seeds before germination and kept under atrazine's influence for 3.5 days in the dark followed by a light period of 8 hours, benzyladenine stimulated the chlorophyll formation, but atrazine inhibited it. This atrazine inhibition of chlorophyll formation was readily overcome by benzyladenine which indicates that atrazine is acting as a competitive inhibitor at the benzyladenine reaction site.

As a new hypothesis, it was investigated to determine whether possible interference of these chemicals on carotenoid or xanthophyll formation in the dark, preceding chlorophyll formation, could account for their effect on chlorophyll content since enhanced carotenoid synthesis could prevent photo-oxidation and photo-destruction of the chlorophyll by removing oxygen from the excited chlorophyll-oxygen complex by the carotenoid epoxide cycle (KRINSKY 1967). The spectra for the carotenoid and xanthophyll fraction from squash cotyledons after benzyladenine treatment significantly increased the absorption peaks at 281, 414, 420, 443, and 472 nm. Atrazine treatment induced significant increase in absorption values only at 281 nm, thus indicating that it is interfering at least in one step of the benzyladenine action on carotenoid or xanthophyll formation (PENNER and WILEY 1972).

A cell division response of tobacco tissue to simazine and atrazine was reported first by JORDAN *et al.* (1966). Growth inhibition of dark-grown tobacco pith occurred when atrazine or simazine at 10^{-3} to $10^{-7}M$ concentrations was added to a nutrient medium containing 0.05 ppm of kinetin: 50% inhibition occurred at $10^{-6}M$. In further tests with soybean tissue, stimulation of growth occurred at less than $10^{-9}M$ atrazine (EBERT and VAN ASSCHE 1969). Stimulation or inhibition of soybean callus growth by triazines was reported at other times (COPPING 1971, COPPING *et al.* 1972, DAVIS and COPPING 1970, OWENS *et al.* 1970). Comparison between these experiments and results are difficult because of varying kinetin and triazine concentrations in the basal medium and because of different kinetin requirements for the different callus types under investigation. In general, high concentrations of the triazines inhibit and low concentrations stimulate callus growth in a kinetin-containing nutrient medium. It seems certain that triazine herbicides change the auxin-cytokinin ratio in the plant tissue, but it remains uncertain how this occurs.

The triazine action affects only a small part of the total spectrum of kinetin's biological activity. Triazines do not replace kinetin and, therefore, do not act as a cytokinin in the plant tissue. Soybean callus did not grow in a nutrient medium containing atrazine but no kinetin (EBERT and VAN ASSCHE 1969). As indicated earlier, senescence biotests with leaf discs of *Rumex, Taraxacum, Avena,* or *Sinapis* showed no effects from atrazine. Biosynthesis of β-cyanin is strongly dependent on the concentration of available kinetin. It cannot be replaced by atrazine (EBERT and VAN ASSCHE 1970). In a leaf expansion test with cotyledons of radish, atrazine had no effect either in the presence or absence of kinetin, although kinetin alone was stimulatory (COPPING *et al.* 1972). However, cotyledons from mustard seedlings were increased significantly in size when the plant was grown in a nutrient medium containing 0.04 ppm of simazine for 18 days (EBERT 1968). The findings from these last two tests again illustrate that the whole plant can undergo changes in metabolism which lead to growth responses but not necessarily correlating to the direct influences of the triazines on auxins and cytokinins as demonstrated in explants.

There are some experimental data on triazine toxicity to plants which do not fit into the general scheme of photosynthesis inhibition. Investigations of simazine's toxicity to pine seedlings led to the conclusion that more than just photosynthesis effects cause phytotoxicity (UHLIG 1966 a) as mentioned in the paper by HARTMANN (1970). The Hill reaction was less disturbed in pine seedlings under 28,000 Lux than under 8,000 Lux, but the plants died first at the higher light intensities.

An attempt has been made to explain this fact by possible influences of simazine on the phytochrome system as a first step involved in causing phytotoxicity. This hypothesis was followed by measuring anthocyanin synthesis in *Oenothera biennis*, which is known to be phytochrome-linked. Simazine at concentrations of 10^{-4} and $10^{-5}M$ exerted a significant reduction of anthocyanin biosynthesis after 16 days of incubation (HARTMANN 1970). The same occurred in the cotyledons of *Sinapis* plants with atrazine treatment in the light, but not under dark conditions (MUELLER 1970).

In germination tests with tobacco seeds exposed for 4 minutes to red light, the germination was increased with or without kinetin. In this test, ametryn behaved like kinetin in sensitizing the light requirement mechanisms of these seeds (COPPING *et al.* 1972). These experiments were extended and the action of the kinetin was compared with that of several *s*-triazines. Kinetin stimulated or inhibited the germination rate of lettuce seed after short exposure to red light depending on the concentration under investigation, but kinetin always stimulated germination in the dark. If there was a direct influence of the triazines on kinetin synthesis or effectiveness, the germination rate should be changed when adding triazines with the well-defined kinetin conditions. In red light, the triazines increased germination, but no reduction of germination occurred at any concentration tested. The triazines did not stimulate germination in the dark. The authors concluded that an effect of the triazines on seed germination is not explainable through a direct effect on kinetin synthesis or effectiveness (PILLAI and DAVIS 1973 a).

Another light-independent phenomenon is the influence of several triazines on the hook opening of cotton hypocotyls. Prometryn and ametryn at concentrations of $10^{-8}M$ stimulated hook opening, whereas atrazine and simazine had negligible effects (DAVIS and PILLAI 1972). The effect of the triazines at $10^{-7}M$ on hook opening decreased in the following order: prometryn > ametryn > prometon > propazine =

kinetin $\geqq$ simazine $>$ cyprazine $=$ atrazine $=$ control. Kinetin was stimulatory in the light and inhibitory in the dark. The triazines in this test behaved like kinetin. IAA had no significant effect, either in the light or in the dark. From these tests, it is clear that neither a direct influence of the triazines on IAA physiology by changing IAA contents in the plants through IAA peroxidase nor a direct influence of the triazines on cytokinins can be the answer to explain the complexity of the influence of the triazines on hormonal reactions in the plants (PILLAI and DAVIS 1973 a and b).

The significance of the triazine action on phytochrome may not be unique, as other photosynthesis inhibitors such as CIPC were also shown to inhibit the phytochrome system (MANN *et al.* 1967). However, it should be considered when explaining some of the physiological effects of the triazines.

In attempting to explain the regulator-like effects of triazine herbicides from all the results, only a few generalities can be drawn. Auxin, as well as cytokinin, metabolism was shown to be involved in the triazine action, but triazines do not seem to work as a plant hormone per se. It seems likely that the auxin-cytokinin ratio undergoes changes in the specific plant tissue leading to the manifold visible effects. How this occurs is not understood at this time, but specific enzymes such as IAA-oxidase could possibly be involved or influence of the triazines on energy-dependent auxin distribution systems in the plant. Cytokinin metabolism was affected by the triazines in some ways, but these cover only a small part of the whole spectrum of cytokinin activity. Gibberellic acid metabolism was not influenced by triazines. Attempts have also been made to explain growth reactions at the protein and RNA level in previous sections of this manuscript. One is left with the obvious conclusion that only through additional study will we elucidate the mode of action of triazines on hormonal control of plant growth.

VI. Seeds

Possible influences of herbicides on the seeds of plants are of major concern, because chemical weed control in agricultural crops should not affect such plant physiological processes as germination, seed formation, or seed quality in a negative way. Since the development of triazine herbicides, their influence on seeds has been investigated from the early stages preceding germination to the formation of new seeds and even to the physiological behavior of these seeds following several generations of plants.

a) Absorption processes

Preceding germination triazines are adsorbed on the seed coat and absorbed by the seed (HASKELL and ROGERS 1962, HOCOMBE 1968,

PHILLIPS *et al.* 1972, RIEDER *et al.* 1970). Two distinct types of herbicide uptake occur, initial accumulation on the seed coat and then uptake directly into the young seedling after cracking. The first step is a physical one and independent of water uptake, in which case the herbicide moves to the seed surface by the mechanism of diffusion. It occurs equally in living or dead, dry or pre-soaked seed (HOCOMBE 1968, RIEDER *et al.* 1970). Only the second step of this two-stage accumulation process depends on the ingress of water into the developing seedling (HOCOMBE 1968). HASKELL and ROGERS (1962) applied ^{14}C-labeled simazine to fully imbided seeds of corn, soybeans, and *Datura stramonium*. After an incubation time of three hours, simazine was found in the seeds of corn and soybeans, but not in the dormant seed of *Datura*. HOCOMBE (1968) studied the absorption of ^{14}C-atrazine on intact and cracked seeds of *Brassica rapa*. Only the cracked seeds took up more water than would be absorbed by imbibition. They also accumulated the greatest amount of atrazine. Most of the herbicide remained on the seed coat, as peeled seeds showed less phytotoxicity in a germination test. The absorption of atrazine into the seeds of soybeans was shown to be strongly dependent on temperature (RIEDER *et al.* 1970, RIEDER and BUCHHOLTZ 1968). Uptake of the herbicide at 30°C was substantially higher than at 10°C. The temperature coefficient Q_{10} was 1.7. This indicates that the herbicide uptake is not only a physical process, but that energy-linked steps may be involved.

The degree of absorption of several herbicides by seeds was shown to vary between different seed species and also between strains of a single species. The total quantity of atrazine absorbed by 11 soybean strains showed large and significant differences between the strains (PHILLIPS *et al.* 1972). It was assumed that the amount of a herbicide absorbed by a given seed would be closely related to the physico-chemical properties of the herbicide with chemical components of seeds such as oil or proteins. The role of oil and protein content of soybean seeds in the absorption process of atrazine and some other herbicides was investigated by PHILLIPS *et al.* (1972). There was a linear correlation between uptake of the herbicide and total oil, and % oil in the seeds, which indicates that oil content is a factor that determines total uptake of the herbicides by the seed. Percent protein was not related to total uptake of herbicides as closely as was total oil and percent oil.

b) Germination

Many investigations concerning the influence of triazines on seed germination have been reported. According to the individual species under study and the herbicide concentration and experimental methods used, varying effects can be demonstrated: inhibition, stimulation, or no effect (ARAI *et al.* 1967, CHAMBERLAIN and ALLEN 1963, CHESALIN and KOVALEVA 1973, COLE and COATS 1973, COPPING 1971, DION 1964/65,

GONZALES-IBANEZ 1964, GROVER 1962, HALL 1967, JORDAN and DAY 1968, KAPPELMAN and BUCHANAN 1968, KASASIAN and PARKER 1971, KOZLOWSKI and KUNTZ 1963, KOZLOWSKI and SASAKI 1968, KOZLOWSKI and TORRIE 1965, LEIDERMANN and KRAMER 1963, LORENZONI 1962, PHILLIPS *et al.* 1972, PILLAI and DAVIS 1973 a and b, SANTELMANN and EVETTS 1971, SASAKI and KOZLOWSKI 1968, SASAKI *et al.* 1968, SINGH *et al.* 1965, SUND and NOMURA 1963, TAS 1961, ZAHN 1966, ZAVITKOVSKI and NEWTON 1964). Only a few of these authors subsequently investigated the influence of triazines on well-defined processes during germination. The viability of the seeds, the germination capacity, and the germination velocity must be taken into consideration. The emergence of the young seedling after germination is quite a different problem.

The viability of seeds can be reduced gradually or totally by soil sterilants. This loss of viability can be shown best with the tetrazolium method (RIEDER and KOCH 1967). KOCH (1967 and 1968) found no effect of simazine on the viability of the weed species *Sinapis arvensis. Veronica persica, Galinsoga parviflora, Alopecurus myosuroides,* and *Vicia hirsuta.* There was a negligible influence of atrazine, simazine, and propazine on seed viability of *Echinochloa crus-galli, Setaria viridis, Setaria glauca, Amaranthus albus,* and *A. caudatus* (FISJUNOV 1970). Likewise, RIEDER and EBERT (1972) found no effect on the crop seeds of *Avena sativa, Hordeum sativum, Secale cereale,* or on the weed seeds of *Sinapis alba* and *Avena fatua.*

The germination capacity, as measured with a germination test in Petri dishes, gives us an answer to a possible influence of a herbicide on seed dormancy, *i.e.*, the shortening of a dormant phase or the induction of a secondary dormancy. Germination capacity was not influenced by atrazine in seeds of *Hordeum sativum, Triticum sativum, Secale cereale,* and *Vicia sativa* (ZELEV *et al.* 1967), but it was reduced by atrazine in *Avena sativa, Avena fatus,* and *Sinapis alba* (RIEDER and EBERT 1972) and enhanced by prometryn in *Foeniculum vulg.* and *Pimpinella anisum* (KAZAKOVA 1968). The reduction in germination capacity can be interpreted as an introduction of a secondary dormancy which was shown to be reversible by washing the seeds with water (RIEDER and EBERT 1972). The results of GROVER (1962) on a reduction of germination capacity of stratified and non-stratified seeds of *Picea glauca* in the presence of simazine, and also those of WAKONIG and ARNASON (1958) from barley, might be interpreted by induction of a secondary dormancy.

In Petri dish germination tests, there was no influence of simazine in the germination velocity of seeds of *Brassica oleracea, Brassica rapa, Raphanus antivus, Trifolium repens, Beta vulgaris, Apium graveolens, Daucus carota, Solanum melongena, Lycopersicum esculentum, Satureja hortensis, Thymus vulgaris, Taraxacum officinale, Lactuca sativa, Cichorium intybus, Asparagus officinalis, Allium cepa, Allium porrum, Phleum pratense, Festuca ovina, Festuca rubra, Lolium perenne,* and *Zea mays* (TAS 1961). SASAKI and KOZLOWSKI (1968) had the same results with *Pinus resinosa* soaked for 24 hours in solutions of simazine, atrazine, and propazine. These results were confirmed under similar experimental conditions (KOZLOWSKI and SASAKI 1968, SASAKI and KOZLOWSKI 1968). There was no effect of atrazine on the germination velocity of seeds of *Hordeum sativum, Triticum sativum, Secale cereale, Avena fatua,* and *Vicia sativa* (RIEDER and EBERT 1972). A series of 33 triazines showed very low activity on the germination of *Orobanche aegyptiaca* and *Striga hermontheca* at concentrations of 0.1, 1, and 10 ppm. Only 100 ppm gave reduction in the germination of both species (KASASIAN and PARKER 1971). PILLAI and DAVIS (1973 a) found that a series of triazines stimulated the germination of lettuce seed given 4 min red light, but they were not stimulatory in the dark. In another investigation, simazine, atrazine, prometryn, and terbutryn did not affect the germination process of 45 weed species (LIEBENOV 1971).

In germination tests with atrazine, RIEDER and EBERT (1972) found an enhancement of the germination velocity in *Hordeum sativum* seeds. As these stimulations only occurred during the first two days, it was assumed that triazines may increase water uptake by the seeds, which regulates the beginning of germination. In an attempt to check this assumption, seeds were soaked in a Carbowax (polyethyleneglycol) solution. Carbowax gives a solution with high osmotic value which retards water uptake by the seed. It was demonstrated that atrazine was able to overcome the inhibition of water uptake caused by Carbowax. This verified the assumption that changes in water uptake were responsible for increases of germination velocity in *Hordeum* seeds by triazines.

Stimulation or retardation of seedling emergence under field conditions seems to be a quite different problem and has a close relationship to influences of triazines on plant physiological systems which direct the growth of the young seedlings. These are most likely plant hormones (EBERT and VAN ASSCHE 1969). The triazine action on the degradation of reserve materials and the control of the processes during germination may be another factor influencing seedling development. Phytase activity was shown to be influenced by atrazine (PENNER 1970). Phytase is the enzyme that hydrolyzes phytin, the storage form of phosphate in seeds, and provides the seedling with orthophosphate and mesoinositol during germination.

Low levels of triazine stimulate seedling growth, while high levels inhibit it. Lethal effects of triazines on seedlings arise from inhibition of photosynthesis. Also α-amylase, ATP-ase and protease activity in germinating wheat and mung bean seeds which were inhibited by GS 14254 contribute to a weaker growth of the seedling (DALVI *et al.* 1972). Under these aspects, the contradictory reports can be explained: LORENZONI (1962) reports of stimulation of seedling emergence by low levels of atrazine, ametryn, and prometryn. Similar effects were observed in 14 different varieties of wheat (CHAMBERLAIN and ALLEN 1963) all of which developed toxicity symptoms after 2 weeks. Stimulating effects of a specific triazine were dependent on a specific line. Prometryn improved the emergence of cotton but retarded the growth of the young seedlings (KAPPELMAN and BUCHANAN 1968). Stimulations in seedling growth of ornamentals were observed by PETOJAN (1966) after the application of atrazine, simazine, and prometryn. Coffee seedlings grew faster after atrazine treatment (GONZALES-IBANEZ 1964). There was little effect of simazine, atrazine, prometryn, and terbutryn on the emergence of 45 weed species (LIEBENOV 1971). Variations in the susceptibility of 6 pigweed (*Amaranthus* sp.) species to prometryn found under greenhouse conditions, but not from field tests (SANTELMANN and EVETTS 1971). Also the results by ABERNATHY and WAX (1971) reflected a susceptibility of different *Amaranthus* species against herbicides in general with atrazine as a representative of the triazines.

Physiological factors which were expected to provide some tolerance to emerging seedlings against triazine toxicity were investigated (WETTASINGHE 1968). Seed weight was examined as a plant physiological tolerance factor. Seeds of swede, kale, mustard, wheat, and pea species were graded into different weight categories and grown under diluted simazine concentration. In all these species, the seedlings emerging from the high-weight seeds tolerated higher amounts of simazine than lightweight seeds. Also the carbohydrate reserve in the seeds was shown to have an influence on herbicide toxicity. Wheat in the one-leaf stage was more tolerant to simazine than mustard in the one-leaf stage (WETTASINGHE 1968). Also breeding work was successful in selecting lines of wheat, rape, and mustard with a higher tolerance against simazine. However, the tolerance factor was not at the level of germination (KARIM and BRADSHAW 1968).

c) Flowering and fruit set

During the development of the plants, the triazine herbicides do not directly affect blooming and fruit set. The influence of triazines on plant hormones was shown to be responsible for shifts in flower formation in corn which are observed occasionally (BOLHAR-NORDENKAMPF 1971). Auxin concentrations in the apex and lateral meristems are mainly responsible for the control of tassel and cob development. In the apex, small amounts of auxin occur, but in the lateral buds high amounts occur. IAA application to the apex leads to the formation of female flowers in the tassels, while decreases in the auxin content in the lateral buds lead to hormonal changes which favor male flowers. All these possibilities for shifting the flower formation in corn by auxin were reported to be observed several times in atrazine-treated corn fields (BOLHAR-NORDENKAMPF 1971). In complementary field tests, the influence of the triazines ametryn, atraton, and atrazine on the male inflorescence was examined on the double cross hybrid corn variety, INRA 258, which tends more to bisexuality than single cross hybrids. Ametryn and atrazine were applied as preemergence and postemergence treatments, and atrazine was applied up to 10 times its recommended rate for postemergence treatments. None of the triazines favored a development of pollen sterility, and on a statistical basis there were no differences in the formation of the tassels of treated and untreated plants (LEUTHOLD 1971). This evidence shows that the triazines are not expected to affect the sexual expression in corn and the observations which have been made may be accounted for as an indirect effect on the plants hormone balance resulting from unusual herbicidal stress. Certainly it cannot be a factor in normal corn production, otherwise the many hybrid corn seed producers who depend on atrazine in their weed control programs year after year would have been forced to stop using it long ago.

d) Seed yield and quality

Seed yield after the usage of recommended rates of triazines correspond to manually weeded controls. Thousand-grain weights were not changed in soybean, castor bean, mustard, coriander (VASIL'EV and YAROSLAVSKAYA 1967), wheat (ZELEV *et al.* 1967), proso-millet (*Panicum milianum*) or cotton seed (SOLYANOVA 1968). The total nitrogen-content was not changed in corn (FINK and FLETCHALL 1967, MEL'NITSCHUK *et al.* 1965, SUSU 1966, SUSU *et al.* 1970), or millet seed (JANISEVSKIJ and GRICEVIC 1967). The nitrate content of the pea seed was not altered by treatment with simazine (RIES *et al.* 1968). Corn quality of pea seed was not influenced by prometryn and simazine (KRASCHENNIKOW and NIKITINA 1968). Chlorazine and propazine increased potassium and phosphorus content in the millet grain (MEL'NITSCHUK *et al.* 1965).

In some cases, subherbicidal concentrations of the triazines increase seed protein. This effect was shown to occur in pea plants in the presence of 0.05*M* simazine grown to maturity under controlled environmental conditions as well as in the field under different climatic conditions (RIES *et al.* 1967 a and b). Ultrastructural studies have shown that the cotyledons of these pea seeds contained more protein bodies and more rough endoplasmatic reticula (WU *et al.* 1972 a). Similar effects were observed in bean seeds (SINGH *et al.* 1972 a).

Higher contents of grain protein were determined in several wheat varieties (MCNEAL *et al.* 1969, RIES *et al.* 1970, SCHWEIZER and RIES 1969 a and b), beans and peas (RIES *et al.* 1968, SALUNKHE and SINGH 1971 a and b, SCHWEIZER and RIES 1969 a, SINGH *et al.* 1972 c), rice (DE DATTA *et al.* 1972, *International Rice Research Institute* 1969/70 and 1970/71, MILLER and MIKKELSEN 1970, VERGARA *et al.* 1970), sorghum (TWEEDY *et al.* 1971), corn (SUSU *et al.* 1970, VOLODIN 1968), lupin (DOROSHENKO and YATSYUK 1968), oats (SCHWEIZER and RIES 1969 a), and sweet corn (SINGH *et al.* 1972 c).

With respect to protein increases in seed, the water-extractable protein fraction was influenced by simazine in edible bean and pea seed

(RIES *et al.* 1967 a, RIES *et al.* 1968). The alcohol fraction of corn proteins contained more protein nitrogen from atrazine treatment (SUSU *et al.* 1970). There was no qualitative change in the protein composition of pea seeds by simazine as determined by acrylamid-gel electrophoresis (RIES *et al.* 1967 a). In another study determination of the total amino acid pattern from soluble nitrogen extractions of pea and sweet corn seed revealed quantitative differences from the controls (SALUNKHE *et al.* 1971 a).

Several parameters with respect to the quality of wheat flour were tested: the dextrin value of the grain, ash, protein and wet gluten contents, dextrin value, Zeleny sedimentation value, gas production, Alveograph, Farinograph, and baking. No significant effects on any of these parameters were noted from the herbicide treatments (MAES and STRYCKERS 1966, ZWATZ and WALTL 1974). In other studies, corn constituents and bread volume were not affected (BUKA and VERTIJ 1967). The effects of simazine on the development of barley were studied by GRZESIUK and co-workers (1973 a, b, and c). They measured starch, protein, and gibberellin levels, and the enzyme activities of α-amylase and protease in barley seed used for brewing. Simazine applied in the field inhibited growth somewhat and the resulting seeds had higher starch and protein contents and protease activity was reduced. In germinating barley, simazine promoted protease during imbibition but inhibited it after 48 hours.

Prometryn did not affect soybean seed constituents (KWITKO 1968, PENNER and MEGGITT 1970, STOLLER *et al.* 1973). Also, cottonseed oil quality from field grown cotton was measured. Composition of myristic, palmitoleic, stearic, oleic, linoleic, and arachidic acids in the oil was not affected by the herbicide treatment (SMITH 1971, WILKINSON and HARDCASTLE 1971 and 1972 a and b). Seasonal edaphic variations caused greater variations to cottonseed oil quality than did herbicides. The mean weight and oil content of seeds were not affected by chlorazine (SOLYANOVA 1968). These results are in agreement with those by MANN and PU (1968) who investigated the inhibition of lipid synthesis by certain herbicides. Atrazine had no effect on the incorporation of radioactivity from ^{14}C-malonic acid into lipids by *Sesbania exaltata* hypocotyls (MANN and PU 1968). Prometryn in 5 cotton varieties had no negative influence on fiber quality such as fiber coarseness and strength (SANTELMANN *et al.* 1966). Simazine and atrazine did not affect the alkaloid content in the fodder lupin (DOROSHENKO and YATSYUK 1968).

In storage organs like potatoes there is no influence on starch content and quality when appropriate dosages and recommended cultural practices are used (BURGHAUSEN 1963 and 1968, JESCHKE 1966, SVRKOTA *et al.* 1968, TRISKIN 1968). At overdosages or under certain environmental conditions, dry matter and starch content were lowered; consequently, tasting quality can be influenced (DIMITROV 1965, HUNNIUS 1967, JESCHKE 1966, ROTH 1963). No adverse effect of simazine was noticed on the bulb production of *Lilium longiflorum* (WATERS 1963). Simazine applied to pome fruit for 2 years did not consistently affect respiration, firmness, color, or the soluble solids of the fruit (CARLSON 1967, CARLSON and RIES 1967). Fruit constituents were not affected in peach (ZAVARZIN and BELJAEVA 1968) or blueberries (WELKER and BROGDON 1968). Ascorbic acid content was increased in raspberries (FREEMAN 1967). Grape quality was not influenced by simazine (LEONARD 1961, TARLAPAN and PANCYR' 1966).

Residue analysis was performed using gas chromatographic methods for all species of seeds where triazines had been used as herbicides (MATTSON *et al.* 1970). In all cases, the results from residue analysis were below the detection limit. The triazine herbicides are not accumulated in seeds (EBERLE 1970), potatoes (EBERLE 1970), or fruit such as apples, black currents, grapes, stone fruit (KARNATZ 1966, MAIER-BODE and BECHTEL 1968, OKULOV and TARNAKINA 1971, ZAVARZIN and BELJAEVA 1968), and pecan nuts (WORLEY and HARMON 1967).

e) Second generation

The plant physiological behavior of the seeds from triazine-treated plants was a topic of several intensive studies where, in general, neither germination capacity nor germination velocity was affected (LIANG *et al.* 1967 and 1969, SANTELMAN *et al.* 1966, VASIL'EV and YAROSLAVSKAYA 1967, ZELEV *et al.* 1967). Rice seeds with a higher protein content as a result of a subherbicidal treatment to the parents were shown to result in higher average % germination and greater seedling vigor (MILLER and MIKKELSEN 1970). Winter wheat seeds which were higher in proteins due to triazine treatment or nitrogen fertilization yielded more content of seed during the next generation (RIES *et al.* 1970, SCHWEIZER and RIES 1969 b). The yield of the second generation was closely associated with the increased protein content of seeds and not with the seed weight. In another study, a planting of oat seeds from simazine-treated plants with a higher protein content yielded, in the second generation, more grain with only little alteration in their protein (SCHWEIZER and RIES 1969 b).

Repeated annual use of atrazine at normal rates on corn had no significant effect on corn yield or several corn yield components when applied for three consecutive years at 3 locations. Corn ear-number/100 plants was not affected nor was corn ear-weight/100 plants (BURNSIDE *et al.* 1969). The productivity of highbush blueberries was not reduced after longterm use of recommended rates of simazine (WELKER and BROGDON 1968).

Following 3 generations of *Arabidopsis thaliana*, which is regarded as the botanical *Drosophila* and quickly reveals genetic effects, gave no indication of morphological or genetic aberrations (LEUTHOLD 1971).

In the available literature, little evidence is given for triazines directly influencing processes affecting seeds from the early steps of germination to seed formation to the seed's physiological behavior in the following generation. Sometimes subherbicidal applications of herbicides produce seeds with a higher protein content, and these seeds when planted will produce better. The higher production is attributed to the higher protein content of the seed and not to the herbicide application. This indirect effect on seeds is not passed to successive generations.

VII. Plant variability in triazine resistance

The successful use of any selective herbicide in a given crop/weed situation is determined by the crop's capacity to remain unharmed while the associated weeds are killed; thus, variability in plant response to herbicides is the major factor that permits the control of certain plants (weeds) and the simultaneous protection of others (e.g., crops).

a) Factors affecting resistance

Plant resistance to triazine treatment varies widely in magnitude and commercial significance. Variability occurs between crop groups (NEGI *et al.* 1964); between genera, e.g., various legumes/terbutryn (ESHEL and RUBIN 1972), fruit trees/simazine and prometryn (TWEEDY and RIES 1966), or gymnosperms/simazine (UHLIG 1968); between species, e.g., *Amaranthus*/atrazine (ABERNATHY and WAX 1971); between varieties, e.g., corn/triazine (ANDERSEN 1964, CHAPLIN and ALBAN 1960), flax/atrazine (ANDERSEN and BEHRENS 1967), grape/simazine (LIDER *et al.* 1966), *Setaria*/atrazine and propazine (OLIVER and SCHREIBER 1971), soybeans/atrazine (ANDERSEN 1969), *Sorghum*/atrazine (CRUZ 1969), sugarcane/atrazine (PENG and YEH 1970); between isogenic lines, e.g., corn/atrazine (EASTIN 1963, 1969, and 1971; EASTIN *et al.* 1964 a and b; HAMILTON 1964 a and b), sorghum hybrids and inbred lines (BURNSIDE 1970); between interspecific hybrids, e.g., sugarcane/ametryn (OSGOOD *et al.* 1969), *Zoysia japonica* cultivars/atrazine (SMITH and ILNICKI 1973), sorghum (*Sorghum bicolor*) and sudangrass (*Sorghum sudanense Piper*) cultivars/atrazine (ROETH 1970); and between individual plants (KARIM and BRADSHAW 1968, WARWICK 1973). Also, the physiological state of the plant plays a role, because with advanced development one generally finds an increase in tolerance (ROBINSON 1974, UHLIG 1966 a and b, HILL and VENGRIS 1968, JANYSKA 1965, SCHUBERT and WEISE 1968).

Studies of photosynthetic inhibition with the use of isolated chloroplast suspensions have shown that inhibition is brought about by triazines whether such chloroplasts are taken from susceptible or resistant species (MORELAND and HILL 1962). Thus it seems that the chloroplast itself does not determine selectivity. Rather, plant response appears to depend on the extent to which the triazine remains in a toxic concentration in the chloroplast. It follows that in plants that are able to survive triazine treatment, a toxic concentration is in some way prevented from accumulating in these organelles.

In theory, there are 4 ways in which the movement of triazines from the soil to the chloroplasts may be slowed down or completely prevented:

(1) Physiological factors such as seed size and rooting habit may reduce triazine uptake (positional selectivity). Also root membrane characteristics could reduce absorption below toxic levels.

(2) Triazine absorption and translocation may be impeded.

(3) The triazine may be partially or completely detoxified on the way to the leaf tissue.

(4) Subsequent triazine movement within the plant may be limited by localized binding or accumulation at inactive sites, compartmentalization.

1. Positional selectivity and other physiological factors.—Positioned selectivity is important in many deep-rooting crops (GYSIN 1962). A toxic concentration of a surface-applied triazine does not reach the roots of these otherwise susceptible crops, but it is able to control weeds that germinate and have their root systems nearer the soil surface.

It is also possible that the capacity to produce long deep roots immediately after germination or to exhibit superior general vigor may permit an individual seedling to show increased resistance to triazine treatment. This may explain in part some of the variability in triazine resistance observed within species (ANDERSEN 1969, MILLER and BOVEY 1969, WARWICK 1973). If the root system of a seedling rapidly extends below the treated zone then because of the progressive reduction in triazine absorption with soil depth, and the steady increase in plant volume, the internal concentration of triazine will become more and more diluted and the seedling may recover from its initial injury.

It has been observed that the largest and heaviest seed in a species sample usually produces the most vigorous seedlings and that seed size also influences the resistance of plants to soil-applied herbicides.

WETTASINGHE (1968) compared the resistance of swede, marrow-stem kale, white mustard, wheat, and peas that had been graded according to seed weight in soil treated with simazine. Although large seed size did not explain the differences in resistance *between* species, within all species tested the seedlings arising from heavier seeds survived higher concentrations of simazine than those from lighter seeds. In addition, the ED_{50} (the dose causing a 50% reduction in shoot weight) was significantly greater for large seeds than for small ones.

In soybeans, ANDERSEN (1969) found that 80% of the variation in atrazine resistance was attributable to seed size, and concluded that the cotyledonary food reserves may determine seedling resistance in this crop. Given that atrazine gradually disperses with time, perhaps large-seeded strains can survive on their reserves until the internal herbicide concentration reaches a non-toxic level and the production of photosynthate is no longer inhibited. Also a vigorous seedling with large cotyledons may have a smaller absorbing surface relative to total volume and hence exhibit less damage after triazine treatment than a smaller seedling. Finally, vigorous seedlings with a large leaf area are well equipped for recovery and subsequent growth when the herbicide concentration within the plant has fallen to a noninhibitory level.

MILLER and BOVEY (1969) found that the degree of resistance to propazine and terbutryn shown by various sorghum varieties appeared to be related to vigor and speed of growth. Also, relations in grasses of high photosynthetic capacity (C-4-plants) and tolerance to atrazine have been the subject of investigations (IMBAMBA and MOSS 1971, IMBAMBA 1970, MOSS 1968 a and b, BOLHAR-NORDENKAMPF 1970).

2. Absorption and translocation.—Work with radioactive simazine has shown that all plants so far tested are able to take up considerable amounts of this herbicide through the roots (CRAFTS 1965, MONTGOMERY and FREED 1961, NEGI *et al.* 1964, SHEETS 1961). Triazine uptake appears to be a passive process related to the concentration of herbicide in solution and the movement of water through the plant; the transport seems to be confined to the apoplast. In fact, this simple mode of translocation contributes to the success of the triazines as herbicides. Considerable uptake and translocation takes place even in resistant plants (DAVIS *et al.* 1959, HAMILTON and MORELAND 1962, SHONE and WOOD 1972 a and b). There is no sign of a barrier to uptake in the roots, therefore, non-absorption/non-translocation cannot be regarded as probable resistance mechanisms. However, the rate of absorption and translocation may account for slight differences in triazine resistance, as found in *Panicum* and *Digitaria* (ROBINSON 1974) and in resistant and susceptible lines of corn (EASTIN 1969, SCHULTZ and TWEEDY 1971).

The tolerance of plants toward triazines may also be influenced by differing environmental conditions. Temperature, light availability, and humidity change the rate of absorption, translocation, and detoxication of the triazine and hence the tolerance of the plant under investigation (ARAI *et al.* 1967, GOSSELINK and STANDIFER 1967, IBARAKI 1967, FIGUEROLA 1970, FURTICK 1969, THOMPSON and SLIFE 1970, HOUSEWORTH and TWEEDY 1971, FIGUEROLA and FURTICK 1972 a and b, KOZLOWSKI *et al.* 1967, ROETH 1970, SHONE and WOOD 1972 a and b, SINGH 1972, SINGH *et al.* 1972, UHLIG 1966 a and b, WU *et al.* 1974). Soil factors, influencing phytotoxicity, were discussed thoroughly at the International Symposium concerning *s*-triazine-soil interactions at Riverside, 1969 (GUNTHER and GUNTHER 1970).

3. Detoxication.—Genuine biochemical resistance is of major importance with regard to the triazines. Work on the chemistry and genetics of such resistance has been done mainly with corn.

Corn possesses at least 2 distinct mechanisms of detoxifying atrazine. These include hydroxylation, which is nonenzymatic and catalyzed by benzoxazinone (HAMILTON 1964 b, HAMILTON and MORELAND 1962, SHIMABUKURO 1968), and enzymatic conjugation processes (SHIMABUKURO *et al.* 1971). A detailed description of these degradation pathways is given in the triazine chapter of KEARNEY and KAUFMAN'S book (1975).

These detoxication reactions are so effective that toxic levels of atrazine (and simazine) do not accumulate in corn even when high rates are used for effective preemergence and postemergence weed control.

Although most varieties of corn are resistant to atrazine and simazine, susceptibility has been found in the Mississippi selection of the inbred line GT 112 (EASTIN 1971, GROGAN *et al.* 1963). This is due to a single recessive gene (GROGAN *et al.* 1963), located on the long arm of chromosome 8 (SCOTT and GROGAN 1969). The discovery of this gene has proved very useful, permitting investigation of the mode of action of the triazines in isogenic lines within a species and susceptibility in another (EASTIN *et al.* 1964 a and b).

Hydroxylation of chlorotriazines in resistant corn is catalyzed by benzoxazinone (BOA) and its derivatives (CASTELFRANCO *et al.* 1961, SHIMABUKURO 1968). ANDERSEN (1964) concludes that there is a relationship between corn resistance to the European corn borer, stalk rot, and triazine herbicides, all of which seem to be dependent on the activity of BOA and its related compounds.

On finding that triazine-susceptible seedlings of wheat and rye contain lower concentrations of BOA derivatives than corn, HAMILTON (1964 a) examined the resistance of recessive segregates from a line of corn containing BOA at a level comparable to these susceptible species. The resistance character behaved as a single recessive gene upon inbreeding and there was clearly a reduction in resistance associated with low BOA content. Nevertheless, these susceptible segregates were relatively resistant to chlorotriazines and did not show the acute phytotoxicity that the author observed in wheat containing a comparable level of BOA.

In fact, the assumption that the rate of hydroxylation is directly correlated with BOA concentration is now recognized to be an over-generalization. One reason for this is that the resulting hydroxytriazines have been found in plants devoid of BOA (HAMILTON 1964 a and b). Moreover, BOA has been detected in susceptible plant species. Clearly, BOA hydrolysis is important; but nevertheless, only one of several mechanisms contributing to the phenomenon of biochemical resistance to triazines.

The chlorotriazines can also be detoxified by cultivated sorghum (SHIMABUKURO 1968). A comparison of atrazine uptake, metabolism, and resistance in sorghum and corn was made by ROETH (1970). Unfortunately, the chlorotriazines can also be detoxified by certain grass weeds. These include the wild cane (THOMPSON 1972 a), *Digitaria, Panicum,* and *Setaria species* (THOMPSON 1972 b, ROBINSON 1974). Triazine degradation has also been reported in *Coix lacrima* and *Imperata cylindria* (HURTER 1966 and 1967 a and b). This explains in part why these grasses are becoming problems.

4. Localized binding/accumulation at inactive sites.—The triazines are strongly adsorbed by organic materials in the soil and this affects their subsequent availability to plants (HAYES 1970). It appears that adsorption may also occur on certain similar materials within the plant so that a lower concentration of triazine reaches the chloroplasts than was initially taken up by the roots. Such localized accumulation may explain, at least in part, the resistance of certain species to triazines. Cotton is moderately resistant to several triazines (HOLSTUN and BINGHAM 1960). DAVIS *et al.* (1959) studied normally glanded and genetically glandless varieties and found that triazines accumulated in the gossypol or lysigenous glands. It has been suggested that the accumulation of simazine (FOY 1962), prometryn (WHITTENBERG 1965), and ipazine (HAMILTON and MORELAND 1963) in such glands is partially responsible for the moderate resistance of cotton to these herbicides. Simazine accumulation has also been observed in cotton roots (SHEETS 1961), and RUBIN and ESHEL (1973) reported the accumulation of terbutryn in the seed coat, roots and gossypol glands. A certain amount of triazine degradation is also known to occur (SHIMABUKURO and SWANSON 1970, SIKKA and DAVIS 1968).

Blackcurrant is moderately resistant to simazine. When blackcurrant, marrow, and barley were grown in a culture solution containing 0.2 ppm labeled simazine for 48 hours, 40% of the total activity was concentrated in the roots of blackcurrant compared with 16 and 12% in the susceptible barley and marrow (SHONE and WOOD 1972 a and b). Autoradiographs of treated plants revealed further accumulation of simazine in the leaf veins of blackcurrant. In contrast, those of susceptible oats and cucumber showed accumulation at the leaf margins, which is where one would expect a herbicide being carried in the transpiration stream and not impeded or diverted en route.

b) The improvement of crop resistance to triazines by artificial selection and plant breeding

Several researchers have suggested that crop selectivity to triazine herbicides or their residues might be improved by exploiting the natural variability that is clearly present in plants, either by searching for varietal differences in triazine tolerance or by altering the genetic structure of the crop by repeated artificial selection for genuine resistance. The attempts that have already been made have yielded interesting results.

1. Flax.—ANDERSEN and BEHRENS (1967) screened 1541 flax strains with atrazine and found that one single plant selection from C.I.719 was relatively resistant. It produced dry matter equal to 33 and 44% of the untreated material in soil treated with 1.12 kg/ha atrazine whereas control varieties Arny and Koto produced 10 and 11%. This atrazine-resistant selection, designated C.I.2484 (COMSTOCK and ANDERSEN 1968) was not, however, resistant to simazine. When C.I.2484 was crossed with susceptible Koto, and the parental varieties, F1, F3, BC_1F_2 (recurrent parent C.I.2484) and BC_2F_2 (recurrent parent Koto) populations evaluated, the atrazine-resistance character was found to behave quantitatively (COM-

STOCK and ANDERSEN 1968). This is contrast to the simple inheritance reported in maize (GROGAN *et al.* 1963). Heritability values were relatively low, suggesting that progress from early generation selection for atrazine resistance in flax would be limited, but that selection in successive generations (F_4, F_5) might yield significant increases.

2. Soybeans.—In a similar assessment of 2,500 strains of soybean, ANDERSEN (1969) found great differences in atrazine response. However, he concluded that from their screening method, based on a relatively short period of exposure to the herbicide, large-seeded strains were selected rather than those with genuine atrazine resistance. The strains producing the most dry matter relative to their controls had very large seeds and those producing the least had very small seeds.

The new techniques of culturing plant cells in suspension cultures and regenerating whole plants from these cells provide a new tool for selectingresistant strains. By these methods a first approach was made by selecting an atrazine resistant soybean tissue from diploid soybean cell suspension cultures (ZENK 1974).

3. Grain sorghum.—WIESE and QUINBY (1969) studied the inheritance of propazine resistance in grain sorghum and compared it with that of 2,4-D. Of 60 varieties screened in 8.96 kg/ha propazine, Martin and Red Kafir emerged as resistant, and Caprock and Pink Kafir as susceptible. Evaluation of F3s from crosses between these varieties indicated that the inheritance of resistance to propazine injury was dominant, although more than one gene seemed to be involved. In contrast, resistance to 2,4-D was clearly dominant and controlled by a single gene.

4. Wheat, oilseed rape, and mustard.—KARIM and BRADSHAW (1968) screened 40 varieties of wheat, 10 of oilseed rape, and 4 of mustard to see, firstly, whether resistance to simazine occurred in these crops, and secondly, whether it could be increased by repeated selection. Rape and mustard are outbreeders, so variability would be expected both within and between varieties. Wheat is by contrast an inbreeder, so that variability in resistance would be mainly between varieties.

Varietal differences in survival and final weight of survivors were found in all three crops but differences in the survival of individual plants were greatest in the two outbreeders. Progenies from the 1% of mustard plants surviving 0.7 kg/ha simazine, which had been thoroughly incorporated into the soil, showed a very marked increase in resistance over the unselected controls. The authors maintained that, given a reasonable amount of genetic variability in the initial populations, it should be possible to select directly and rapidly for herbicide resistance within other existing crop varieties. Of course, whether or not such resistance would be correlated with unacceptable characters could not be forecast.

The feasibility of selecting for simazine resistance in oilseed rape was investigated further by WARWICK (1973). A significant change was achieved in one line after 3 cycles of selection, and this resistance also showed up in field trials. It appeared to be controlled by polygenes as a number of factors related to seed size, general vigor, and rates of photosynthesis, translocation and detoxication of simazine contributed to its increase.

An extensive source of variation which could provide a broad genetic base for the improvement of simazine resistance was found in other members of the *Brassica* group also.

In heavy soils, many *Brassica* are protected by deep planting and soil adsorption of triazine. However, on light soils, there is the risk of damage if heavy rain follows application and carries the triazine to the crop roots. For this reason, dose rates have to be kept low and unsatisfactory weed control is frequently observed in such situations. Selection for enhanced triazine resistance would permit more efficient weed control and lessen the probability of crop damage (Warwick 1973).

A suitable level of increased resistance has been achieved by utilizing natural varietal differences (Andersen 1964, Wiese and Quinby 1969) or by selecting for individuals with superior resistance (Andersen and Behrens 1967, Comstock and Andersen 1968, Karim and Bradshaw 1968, Pinthus and Eshel 1973, Warwick 1973). Pinthus *et al.* (1972) investigated the feasibility of improving herbicide resistance by means of induced mutations. From wheat seeds treated with ethyl methane sulphonate, mutants were successfully selected with increased seedling resistance to terbutryn. All these studies have aroused interest in the possibility of incorporating selection for resistance to existing herbicides into future plant breeding programs.

c) The natural evolution of triazine resistance in weeds

Since no one herbicide is effective against all weed species, the presence of so-called resistance or problem weeds is almost an inevitable consequence of herbicide usage. Such resistant weeds have never shown any susceptibility to the relevant herbicides and the resulting weed flora which evolves is due to the gradual elimination of susceptible species. Triazine-resistant weeds include various tropical and sub-tropical grasses (Hurter 1966, Thompson 1972 a and b, Thompson *et al.* 1971). There is now some evidence that resistant strains are being selected naturally from susceptible weed populations particularly in crops which receive repeated applications of the same herbicide (Radosevich 1973 a and b, Radosevich and Appleby 1973 a and b, Ryan 1970 a and b, Daris 1968).

Lack of control of common groundsel (*Senecio vulgaris*) was first observed in 1968 in a nursery where simazine or atrazine had been used once or twice annually for ten years (Ryan 1970 a and b). Seedlings from that location were not affected by preemergence applications of 17.92 kg/ha and were designated resistant (*R* biotype). However, seedlings from a source where triazines had not been used continuously were completely controlled by 1.12 or 2.24 kg/ha simazine or atrazine and were labeled susceptible (S biotype). Postemergence application of 1.12,

2.24, and 4.48 kg/ha atrazine failed to control the *R* biotype, but killed S plants. In contrast, preemergence applications of dichlobenil, chloroxuron, fluometuron, and norea were equally effective against both biotypes. S plants were killed by 0.5 ppm of atrazine or simazine, 1.0 ppm of GS-14254 or of prometon, and 4.0 ppm of prometryn, whereas the *R* biotype was unaffected by the highest rate (30 ppm) of atrazine used (RADOSEVICH and APPLEBY 1973 a, b, and c).

In a further study, (RADOSEVICH and APPLEBY 1973 c) photosynthesis was completely stopped by simazine in *S* plants, whereas *R* plants were unaffected. When simazine was removed after 24 hours, photosynthesis resumed in the *S* biotype. Both biotypes absorbed the herbicide equally well and were unable to metabolize it during 96 hours. In fact, 80 to 90% of ^{14}C activity was located in the chloroform-soluble fraction of the foliage of both biotypes and was identified as simazine. This suggests that degradation does not play an important role in the resistance of common groundsel to chlorotriazines. The authors suggested that perhaps the *R* biotype forms a weakly bonded simazine conjugate which is inactive in intact plants, but which releases simazine during extraction. Another factor may be the rate of photosynthesis, which is significantly higher in the *R* than in the *S* biotype in the absence of simazine.

In the United Kingdom, repeated applications of simazine are used to control weeds in many fruit crops. HOLLIDAY and PUTWAIN (1974), investigating the possibility of simazine-resistant weeds developing in such crops, have found that after 11 years of continuous application of simazine the natural resistance of groundsel has increased slightly. Populations having received 11 years of regular treatment resulted in 13% survival when treated with 0.6 kg/ha simazine thoroughly incorporated into the soil compared with 3% survival of populations that had never received simazine treatment.

By examining carefully the conditions that have brought about the resistance observed in groundsel (RYAN 1970 a, HOLLIDAY and PUTWAIN 1974), one can make certain predictions as to the likelihood of triazine resistance developing in other weeds.

The rapid selection of such a character requires a heavy and repeated selection pressure, *i.e.*, the isolation of selected plants to prevent outbreeding with unselected material and subsequent dilution of loss of the "resistance" character (HARPER 1956). Thus, in order to prevent weeds from evolving herbicide resistance, conditions favoring its selection should be avoided by practicing herbicide rotation, or employing herbicide mixtures.

In artificial selection procedures, a very large population may be screened at a given time and all plants subjected to the same screening dose. In contrast, natural screening pressures touch only a small part of the weed population at a given time due to the large reservoir of unselected dormant seeds and unselected plants growing at the edge of the area treated with herbicide. Individuals from this reservoir will readily interbreed with selected plants, thus maintaining a relatively constant level of resistance in the population as a whole. This, in turn, reduces the probability of a crop plant artificially selected for triazine resistance passing on its resistance to genetically related weeds. Further exploitation of the natural variability in plant resistance to triazines seems a worthwhile venture.

VIII. Water balance

Since SMITH and BUCHHOLTZ (1962) first reported on inhibition of plant transpiration by atrazine, many published reports have included the same observation for several triazine herbicides on a wide variety of plants, including corn, oats, wheat, red pine, white spruce, soybeans, rice, vetch, peas, cotton, peanuts, lima beans, snap beans, sugarbeets, barley, tobacco, tomatoes, sunflower, squash, cucumbers, and okra (CINCOVIC *et al.* 1967, CLAUSEN and KOZLOWSKI 1964 and 1966 b, GILL 1968, GLABISZEWSKI *et al.* 1966, GRAHAM 1967, GRAHAM and BUCHHOLTZ 1968, HUMBURG 1971, IMBAMBA and MOSS 1971, OORSCHOT 1965 and 1970 b, SIKKA *et al.* 1964, SIVADJIAN and RICARDO 1965, SMITH and BUCHHOLTZ 1962 and 1964, SMITH and ILNICKI 1972 b, WAGGONER and ZELITCH 1965, WALKER and ZELITCH 1963, WILLIS and DAVIS 1962, WILLS *et al.* 1963, ZELITCH 1969). In nearly all reports, the extent of transpiration reduction, which varies from 0 to 70%, was related directly to the herbicidal susceptibility of the particular plant species. For example, corn which is not very susceptible to atrazine was affected less than soybeans, one of the most susceptible crops. Indeed, PALLAS and BERTRAND (1967) found no significant effect on transpiration or stomatal activity at all in corn under field conditions. They explained their results with respect to the positive findings from nutrient culture experiments (SMITH and BUCHHOLTZ 1962 and 1964, WILLS *et al.* 1963) by differences in atrazine uptake and CO_2 levels which occurred between the 2 experimental systems. Under most field conditions, the corn plant is unable to take up atrazine fast enough to affect photosynthesis or any other mechanisms controlling stomatal movement since corn detoxifies atrazine so rapidly. In nutrient cultures, the detoxifying mechanism is probably overloaded to the point where some herbicidal action is taking place.

In all the reports of transpiration being reduced by the triazines, partial or complete stomatal closure occurred. The effect must be a result of the peristomatic cells losing turgor in relation to the surrounding epidermal cells. This can happen if there is an effect either on the permeability of the cell membranes or on the established ionic gradient contributing to the cell's osmotic pressure. Since there is no evidence that the *s*-triazines can affect membrane permeability, we are left with the latter explanation that they reduce the ionic gradient either directly or indirectly in the peristomatic cells. As evidence against permeability effects, we note that the nonherbicidal triazine, hydroxyatrazine, has no effect on transpiration (GRAHAM 1967). Also, the work of KRIHNING (1965) on the effects of simazine on plasmolysis and deplasmolysis of potato epidermal cells definitely indicates that membrane permeability effects cannot explain decreased transpiration. She observed no simazine effect on glucose-induced plasmolysis with only a slight increase in deplasmolysis time, which may not have been significant.

The ability of the triazine herbicides to inhibit photosynthesis is very likely the initial cause which leads to the effect on transpiration. IMBAMBA and MOSS (1971) and IVAN OORSCHOT (1970 a and b) showed very clearly that CO_2 uptake is reduced more than transpiration in the light. WILLS *et al.* (1963) found the effect with plants in continuous light, but not with plants in continuous darkness. HUMBURG (1971) observed that transpiration rates during the photoperiod were suppressed by $1 \times 10^{-6} M$ atrazine, but they were increased over the control during the nyctoperiod. Besides depriving the cell of energy, inhibiting photosynthesis would also raise the CO_2 concentration of the cells. The effect of high concentrations of CO_2 causing stomata closure in the light is well known. HEATH (1948) claimed that the light response showing stomatal movement (opening) may be entirely due to the removal of CO_2 by photosynthesis. Other reports say that the light response is due to the production of osmotically active molecules by the photosynthetic machinery. Increased CO_2 concentrations inhibit the glycolate-glyoxylate cycle which oxidizes the reduced nucleotide NADPH which, in turn, leads to the synthesis of the high-energy molecule adenosine triphosphate (ATP) by noncyclic photophosphorylation (WAGGONER and ZELITCH 1965). The energy of ATP needed to drive an ion "pump" of the guard cells would not be available to maintain their osmotic potential and the stomata would close under conditions of inhibited photosynthesis.

Current evidence (FISCHER 1968 a and b, FISCHER 1971, GRAHAM and ULRICH 1972, HUMBLE and HSIAO 1969, SAWHNEY and ZELICH 1969, TURNER 1972, ZELITCH 1969) suggests that potassium is the important osmotic agent which is pumped into the guard cells at the expense of energy from ATP. Using the biological desiccant, fusicoccin, TURNER (1972) demonstrated that it stimulated the uptake of potassium by the guard cells and caused opening of the stomata. He also noted that oxidative phosphorylation, noncyclic photophosphorylation, and cyclic phosphorylation could all serve as energy sources to drive the potassium ion "pump." Since the guard cells of active stomata have chloroplasts and the surrounding epidermal cells do not, light energy can be converted to mechanical energy according to this ion pump scheme in stomata and open against the resistance of the surrounding cells. If photosynthesis is partially or completely inhibited by triazines, we would expect the stomata to remain closed or to partially open.

Whether or not the triazines can influence transpiration in the dark may still be an open question. From most evidence, we would conclude that they have no effect (WILLS and DAVIS 1962, WILLS *et al.* 1963), but as mentioned earlier, HUMBURG (1971) noted a marked increase in transpiration during the dark period. Also, GRAHAM (1967) observed that although atrazine at 4.5 × $10^{-6}M$ reduced overall transpiration of soybeans, it accentuated the transpiration rhythm which was inherent in the plants prior to treatment. Unfortunately, the role of rhythms in stomatal behavior is not very well understood, but they apparently do contribute to night opening which has been observed in some plants (MEIDNER and MANSFIELD 1965). Another possible explanation for triazine-enhanced transpiration in the dark is the possibility that the treated leaves have accumulated a higher level of nitrate ions than normal. (This aspect is discussed in another section.) The work of BUCZEK (1967) would suggest that nitrate ions can stimulate transpiration. This effect would be noticed in the dark but would be rapidly overcome in the light by the much more important effect of retarded photosynthesis.

From all evidence, the triazines seem certain to affect indirectly transpiration in the light and the water balance of plants as a result of inhibiting photosynthesis. This conclusion can be reached no matter which hypothesis of stomatal control is considered. Energy supplies needed to drive ion pumps are reduced. The level of CO_2 in the stomata increases, since it is not being used in photosynthesis. And the amount of small, osmotically active molecules produced in the guard cells as a result of photosynthesis, which included glycolate, is reduced.

In most cases where transpiration has been reduced by triazines to any extent, there is an accompanying decrease in the treated plants' rate of weight gain compared to a control plant. Additions of exogenous carbohydrates will allow treated plants to gain in dry weight while their transpiration rates remain inhibited (GRAHAM and BUCHHOLTZ 1968). Because of this discrepancy, it does not seem feasible ever to use the triazines as antitranspirants in field crops. Susceptible plants will be killed, while resistant plants will detoxify them so quickly that any reduction in transpiration will be of no consequence. We can only conceive of one situation where such antitranspirants may be useful. If a crop is under a drought stress, it may be advantageous to sacrifice some of its photosynthetic capacity rather than have it perish. However, the treatment would have to be at a low subherbicidal level but still high enough to retard partially photosynthesis. Because the proper level of treatment would depend upon the particular plant species and all of its environmental factors, this approach to relieving drought stress seems very risky.

Acknowledgements

We should like to extend our thanks to Prof. A. Trebst, Bochum, and Dr. D. Moreland, Raleigh, for their criticism and advice in the chapter on photosynthesis and to Dr. Moreland for reading the chapter on respiration. We should also like to thank Mr. A. L. Meier, CIBA-GEIGY Ltd., Basle for his help in preparing the translation.

Summary

The literature summary presented endeavors to summarize and collate the direct and indirect effects of the triazine herbicides on the plant metabolism which have been observed to date. The most obvious activity of the triazines is, of course, the inhibition of photosynthesis. Triazines, like phenylureas, carbamates, and acylanilides, are well-known potent inhibitors of the electron transfer process in photosynthesis with the inhibition site located in the area of the second light reaction. The herbicide activity of the triazines is directly related to their intervention in photosynthesis. Plant physiological processes which are connected indirectly with photosynthesis as stomatal movement, transpiration, ion movement, and many other energy-dependent reactions are influenced also.

Growth stimulation and inhibition of plants treated with triazines have been the subject of extensive agronomic, physiological, and biochemical research. Plant hormones, especially auxin and cytokinin, were altered by the triazines. On a molecular basis, however, no adequate explanation can be given for this phenomenon to date. The triazine's effect on the absorption of minerals is often linked with influences on growth; consequently, the scientific interest throughout the world has centered on the effects of the triazines on the absorption of nitrogen and the biosynthesis of proteins. The question as to whether the increase of ion absorption is due to the stimulation of active biochemical processes or to effects on root growth must be left for further studies. It is certain though that the triazines change the ion balance in plants, and, consequently, this can have a most important effect on plant metabolism leading to influences on RNA-, enzyme-, and protein synthesis. On the other hand, the triazines show no activity on DNA synthesis or composition. Physiological processes in seeds once they are formed are generally not influenced by the triazine herbicides. Finally, a review of the plant physiological factors which contribute to the selective activity of the triazines in the various plant species is given.

References

Abernathy, J. R., and L. M. Wax: The susceptibility of seven *Amaranthus* species to several herbicides. Res. Rept. N. Central Weed Control Conf. **28**, 67 (1971).

Adams, R. S.: Phosphorus fertilization and the phytotoxicity of simazine. Weeds **13**, 113 (1965).

——, and H. J. Otto: Phosphate fertility and the herbicidal activity of simazine and atrazine. Abstr. Meeting Weed Soc. Amer., p. 73 (1964).

——, and C. N. Sun: Effects of the phosphorus-manganese-atrazine interaction in soybean plants. J. Agr. Food Chem. **19**, 325 (1971).

——, and W. G. Espinoza: Effect of phosphorus and atrazine on mineral composition of soybeans. J. Agr. Food Chem. **17**, 818 (1969).

Adsuar, J.: Deleterious effect of simazine on mosaic-infected sugarcane. J. Agr. Univ. Puerto Rico **45**, 191 (1961).

Agha, J. T., and J. L. Anderson: Effect of simazine upon pigments, soluble solids, and some oxidative enzymes of apple and cherry. Proc. W. Weed Control Conf. **21**, 44 (1967).

AKOBUNDU, I. O., R. D. SWEET, W. B. DUKE, and P. L. MINOTTI: Synergistic effect of atrazine and alachlor combinations on Japanese millet. Abstr. Meeting Weed Sci. Soc. Amer., 96 (1972).

ALABUSHEV, V. A.: Effect of various herbicides on the protein N content on crop seeds on chernozem and sod-podzol soils. Nauchn. Dokl. Vysshei Shkely Biol. Nauki **4**, 163 (1965); through Chem. Abstr. **65**, 1319 (1966).

ALLEN, W. S., and R. D. PALMER: The mode of action of simazine in barley. Weeds **11**, 27 (1963).

ALLISON, D. W., and R. A. PETERS: Influence of simazine on crude protein and cellulose content and yield of forage grasses. Agron. J. **62**, 246 (1970).

ALLOTT, D. J.: The tolerance of newly planted strawberries to certain soil-acting herbicides. Proc. Brit. Weed Control Conf., p. 881 (1968).

ANDERSEN, K. J., E. G. LEIGHTY, M. T. TAKAHASHI, and A. J. CUTALA: Evaluation of herbicides for possible mutagenic properties. Evaluation of bensulide for mutagenic properties in microbial test system. J. Agr. Food Chem. **20**, 649 (1972).

ANDERSEN, R. N.: Differential response of corn inbreds to simazine and atrazine. Weeds **12**, 60 (1964).

— A search for atrazine resistance in soybeans. Abstr. Meeting Weed Soc. Amer., p. 157 (1969).

—, and R. BEHRENS: A search for atrazine resistance in flax (*Linum usitatissimum* L.). Weeds **15**, 85 (1967).

ARAI, M., S. FURUYA, and M. MIYAHARA: Studies on the mode of action of s-triazine herbicides. 1. Differences among the derivatives of s-triazine in phytotoxicity for rice plant and barnyard grass (*Echinochloa crus-galli*) at the germinating stage. 2. Influences of temperature on the phytotoxicity of methylmercapto substituted triazines for rice plant and barnyard grass. Proc. Crop Sci. Soc. Japan **36**, 310 and 315 (1967).

ARMSTRONG, A. M., and G. W. ANDERSON: Nitrate reductase activity of soybean leaves as affected by atrazine and soybean herbicides. Res. Rept. Can. Weed Comm. E. Sect. **16**, 68 (1971).

Asahi Chem. Ind. Co. Ltd.: JA 7102377. Priority date: 6.7.1967.

ASHTON, F. M.: The effect of atrazine on the metabolism of C^{14}- sucrose in excised bean leaves in the dark. Plant Physiol. **35**, Suppl. XXVIII (1960).

— Action spectra of atrazine injury. Plant Physiol. **37**, Suppl. XXV (1962).

— Relationship between light and toxicity symptoms caused by atrazine and monuron. Weeds **13**, 164 (1965).

—, and T. BISALPUTRA: Effect of atrazine on *Chlorella*. Plant Physiol. **39**, Suppl. XXXIII (1964).

— —, and E. B. RISLEY: Effect of atrazine on *Chlorella vulgaris*. Amer. J. Bot. **53**, 217 (1966).

—, E. M. GIFFORD, and T. BISALPUTRA: Structural changes in *Phaseolus vulgaris* induced by atrazine. 1. Histological changes. Bot. Gaz. **124**, 329 (1963 a).

— — — Structural changes in *Phaseolus vulgaris* induced by atrazine. 2. Effects on fine structure of chloroplasts. Bot. Gaz. **124**, 336 (1963 b).

—, and D. PENNER: Effect of several herbicides on the proteolytic activity in squash. Abstr. Meeting Weed Soc. Amer., p. 48 (1966).

—, and E. G. URIBE: Effect of atrazine on sucrose-C^{14} and serine-C^{14} metabolism. Weeds **10**, 295 (1962).

—, G. ZWEIG, and C. W. MASON: The effect of certain triazines on $C^{14}O_2$-fixation in red kidney beans. Weeds **8**, 448 (1960).

ASLAM, M., and R. C. HUFFACKER: Effect of DCMU, simazine and atrazine on nitrate reductase activity in *Hordeum vulgare* in vitro and in vivo. Physiol. Plant **28**, 400 (1973).

AVRON, M.: Mechanism of photoinduced electron transport in isolated chloroplasts. In D. R. Sanadi (ed.): Current topics in bioenergetics, vol. **II**, p. 1. New York: Academic Press (1966).

BAGAEV, V. B., and L. I. KOBAZEVA: Effect of mineral nutrients on sensitivity of corn towards atrazine. Dokl. TSKhA *109*, 207 (1965); through Chem. Abstr. **65**, 16,000 (1966).

BARANOWSKI, R. J. BORS, M. PLOSZYNSKI, and H. ZURAWSKI: Influence of some herbicides on the accumulation of phosphorus-32 in the aerial parts of oats. Pamiet. Pulawski **28**, 63 (1967); through Chem. Abstr. **68**, 113,540 (1968).

BARTH, A., and H.-J. MICHEL: Ueber Wirkungsmechanismen und Substituenteneinflüsse von Herbiziden des Harnstoff-, Carbamat-, Amid- und Triazintyps. Tagungsber. Deut. Akad. Landwirtschaftswiss. Berlin, Nr. **109**, p. 41 (1968).

—— —— Wirkungsmechanismen and Substituenteneinfluss von Herbiziden des Harnstoff-, Carbarmat-, Amid- und Triazintyps. Pharmazie **24**, 11 (1969).

BARTLEY, C.: Simazine and related triazines as herbicides. Agr. Chem. **12**, 34 and 113 (1957).

BASKAKOV, YU A., YU V. SHCHEGLOV, N. A. VOLODARSKAYA, and I. A. MEL'NIKOVA: Herbicidal derivatives of hydroxylamine. 3. The herbicidal activity of 2-chloro-4-[alkyl] dialkylamino-6-methoxyamino-1,3,5-triazines. Agrokhimiya **10**, 135 (1965); through Weed Abstr. **16**, 312 (1967).

BASTIN, R., A. BIENFAIT, G. VAN ROEY, and W. DE CAT: Effect of urea and triazine derivatives on growth and nitrogen incorporation of winter barley and winter wheat. Meded. Rijksfac. Landbouwwetensch., Gent **33**, 1391 (1968).

——, G. VAN ROEY, and W. DE CAT: Action de dérivés d'urée et de triazines sur la croissance et le rendement en azote d'un escourgeon et d'un froment d'hiver (2ème partie). Meded. Fac. Landbouwwet., Rijksuniv. Gent **35**, 999 (1970).

Battelle Development Corp., Columbus, Ohio: US 3567424. Priority date: 8.2.1968.

BEREZOVSKII, M. Y., V. A. KALININ, and R. P. RAZHAEVA: Interactions of 2,4-D and Atrazine. Dokl. TSKhA No. **115**, 217 (1965); through Chem. Astr. **65**, 6,213 (1966).

BISHOP, N. I. Inhibition of the oxygen-evolving system of photosynthesis by aminotriazines. Biochim. Biophys. Acta **57**, 186 (1962).

BISWAS, P. K., and D. D. HEMPHILL: Role of growth regulators in the uptake and metabolism of s-triazine herbicide by tea leaves. Nature **207**, 215 (1965).

BOEGER, P.: University of Konstanz (Germany), personal communication (1974).

BOIKO, V. S., and V. P. PETELKO: The effect of fertilizers on the sensitivity of weeds to simazine. Khimiya sel'-Khoz. **4**, 46 (1966); through Landwirt. Zentralbl. 2-67/05-1057.

BOLHAR-NORDENKAMPF, H.: Die Wirkung Von Atrazin auf den pflanzlichen Gasstoffwechsel. Biochem. Physiol. Pflanzen **161**, 342 (1970).

—— Neuere Untersuchungen über die Einflüsse von verschiedenen Herbiziden auf den pflanzlichen Stoffwechsel. Ber. deut. bot. Ges. **84**, 525 (1971).

BRAGINSKI, L. P. V. G. GRIN, S. V. KOSTENKO, V. V. LAKHIN, and L. V. SURKOVA: Monuron and simazine as algicides. Tr. Vses. Gidrobiol. Obshchestva, Akad. Nauk. SSSR **14**, 52 (1963); through Chem. Abstr. **61**, 15,280 (1964).

BRANDES, W., and R. HEITEFUSS: Nebenwirkungen von Herbiziden auf *Erysiphe graminis* and *Cercosporella herpotrichoides* an Weizen. I. Der Befall der Pflanze unter dem Einfluss der Herbizide. Phytopath. Z. **71**, 357 (1971 a).

—— —— Nebenwirkung von Herbiziden auf *Erysiphe graminis* und *Cercosporella herpotrichoides* an Weizen. II. Physiologische und biochemische Ursachen des veränderten Befalls der Pflanze. Phytopath. Z. **72**, 34 (1971 b).

BRANTON, D.: Quantasomes: The power units of photosynthesis. New Scientist **40**, 19 (1968).

BRENCHLEY, R. G.: The response of tomatoes to atrazine as affected by magnesium and photoperiod. Diss. Abstr. **29**, 3568B (1969).

——, and A. P. APPLEBY: The influence of magnesium on the toxicity of atrazine. Abstr. Meeting Weed Soc. Amer., p. 52 (1967).

—— —— Effect of magnesium and photoperiod on atrazine toxicity to tomatoes. Weed Sci. **19**, 524 (1971).

BRIAN, R. C.: Electrons and the mode of action of certain herbicides: Chem. Ind. **48,** 1955 (1965).

BRUINSMA, J.: Effects of pesticidal treatments on the chlorophyll content of plant parts. Residue Reviews **10,** 1 (1965).

BRUNETTI, N., G. PICCIURRO, and L. FERRANDI: Increase in the absorption of NO_3-(^{15}N) in seedlings of *Triticum durum* by the action of 2-chloro-4,6-bisethylamino-s-triazine (simazine). Agrochimica **16,** 33 (1971/72).

BUCZEK, D.: Metabolic aspects of transpiration of leaves. The effect of the reduction of nitrates on the process of transpiration in *Lycopersicum esculentum* and *Chenopodium album,* Acta Soc. Botanicorum Pol. **36,** 379 (1967).

BUDOI, G., and S. K. RIES: Absorption of radioactive simazine, increased protein content due to atrazine application to rye plants. Studii Cerc. Biochim. **11,** 371 (1968).

BUECHEL, K. H.: Mechanisms of action and structure-activity relations of herbicides that inhibit photosynthesis. Pest. Sci. **3,** 89 (1972).

BUKA, A. J., and S. A. VERTIJ: (Der Einfluss auf den Ertrag und die Backeigenschaften von Winterweizen). Trudi Charkovskogo SCHI im. V. V. Dokucaeva **69,** 29 (1967); through Landwirt. Zentralbl. 2-69/05-1147.

BURGHAUSEN, R.: Unkrautbekämpfungsversuche mit Triazinderivaten im Kartoffelbau. Albrecht-Thaer-Archiv **7,** 645 (1963).

—— Probleme des Herbizideinsatzes im Kartoffelbau. Europ. Potato J **11,** 3 (1968).

BURNSIDE, O. C. Tolerance of selected sorghum hybrids and inbred lines of atrazine. Res. Rept. N. Central Weed Control Conf., p. 64 (1970).

—— Effect of herbicides on seedling emergence force. Weed Sci. **19,** 182 (1971).

——, G. A. WICKS, and C. R. FENSTER: Effect of repeated annual use of atrazine on corn. Agron. J. **61,** 297 (1969).

BUSCHBECK, R., Ch. BELLING, and A. JENDE: Ueber den Einfluss der Herbizide Chlorpropham und Simazine auf die Peroxydase-Aktivität von Brassicaceen-Keimpflanzen. Wiss. Z. päd. Inst. Güstrow **9,** 29 (1971).

——, and U. KRISPIN: Ueber Wirkungen des Herbizids Atrazine auf den Atmungsstoffwechsel von Brassicaceen-Keimpflanzen. Wiss. Z. päd. Inst. Güstrow **9,** 23 (1971).

BUSH, P. B., and S. K. RIES: Effect of atrazine on elongation of the embryonic axis of red kidney bean. Weed Sci. **22,** 227 (1974).

CAMPBELL, W. F., and J. O. EVANS: Influence of some new triazine herbicides on wheat and alfalfa chloroplast ultrastructure. Abst. Meeting Weed Sci. Soc. Amer. p. 53 (1971).

CAMPER, N. D., and D. E. MORELAND: Correlations between acidity of substituted Phenylamides and inhibition of the Hill reaction. Biochim. Biophys. Acta **94,** 383 (1965).

—— —— Sorption of substituted phenylamides onto bovine serum albumin. Weed Sci. **19,** 269 (1971).

CARLSON, D. B.: The influence of simazine on the quality and composition of pome fruit. Thesis, Michigan State Univ., East Lansing (1967); through Chem. Abstr. **69,** 85,625 (1968).

——, S. K. RIES: Effect of simazine on quality and composition of tree fruit. Abstr. Meeting Weed Soc. Amer., p. 16 (1967).

CASELEY, J. C., and L. C. LUCKWILL: The effects of some residual herbicides on soil nitrifying bacteria. Rep. Agr. Hort. Res. Station, Univ. Bristol, p. 78 (1965).

CASTELFRANCO, P., C. L. FOY, and D. B. DEUTSCH: Non-enzymatic detoxification of 2-chloro-4,6-bis(ethylamino)-s-triazine (simazine) by extracts of *Zea mays.* Weeds **9,** 580 (1961).

CHAMBERLAIN, E. W., and C. P. ALLEN: Study to evaluate the tolerance of several winter wheat varieties to three triazine compounds. Res. Prog. Rept. W. Weed Control Conf., p. 91 (1963).

CHANCE, B., and G. R. WILLIAMS: The respiratory chain and oxidative phosphorylation. Adv. Enzymol. **17,** 65 (1956).

CHAPLIN, D., and E. K. ALBAN: Varietal response of sweet corn to simazine. Proc. N. Central Weed Control Conf. **17,** 52 (1960).

CHEN, T. M., and S. K. RIES: Effect of simazine on nitrate content, NR activity and free ammonia in oats. Abstr. Meeting Weed Soc. Amer., p. 31 (1968).

CHENIAE, G. M.: Photosystem II and O_2 evolution. Ann. Rev. Plant Physiol. **21,** 467 (1970).

——, and I. F. MARTIN: Site of Mn function in photosynthesis. Biochim. Biophys. Acta **153,** 819 (1968).

CHESALIN, G. A., and T. K. KOVALEVA: (Der Einfluss chemischer Präparate auf die Keimung von Unkrautsamen.) Chim. sel'skom Choz., Moskva **11,** 41 (1973); through Landwirt. Zentralbl. 2-73/11-0928.

——, and A. A. TIMOFEEVA: (Atrazine distribution in sunflower plants and its effect on nitrogen metabolism.) Trudy, Vsesoyuznyi nauchnoissledovatel'skii Institut Udobrenii i Agropochvovedeniya **51,** 195 (1971); through Weed Abstr. **22,** 000 (1972).

—— —— The effect of atrazine on some characteristics of metabolism in sensitive plants. Agrokhimiya **5,** 108 (1968); through Chem. Abstr. **70,** 27,833 (1969).

CHIESURA, F. R., and G. G. LORENZONI: The effect of temperature on young maize plants treated with simazine. Giorn. Bot. Ital. **71,** 638 (1964).

CHMIEL, H., and S. K. RIES: The effect of simazine concentration on nitrate reductase levels in several plant species. Abstr. Meeting Weed Soc. Amer., p. 60 (1967).

CHODOVA, D., and J. ZEMANEK: Effect of the herbicides MCPA and simazine on the respiration rate and content of glycides and nitrogen in bindweed (*Convolvulus arvensis* L.). Biol. Plant. (Praha) **13,** 234 (1971).

CINCOVIC, T., M. KOJIC, and B. SINZAR: The influence of simazine and atrazine on the water balance of wheat. Abstr. 6th Internat. Congress Pl. Prot., Vienna, p. 394 (1967).

CLAUSEN, J. J., and T. T. KOZLOWSKI: Water relations in nursery seedlings and transplants treated with herbicides. Wisconsin, Univ. Forestry Res. Notes No. **111** (1964); through Chem. Abstr. **65,** 17,620 (1966 a).

—— —— Effect of atrazine on water relations of needles of white spruce nursery stock. Forest Sci. **12,** 338 (1966 b).

COIC, Y., and C. LESAINT: Recherches rus l'action physiologique de la simazine. Comp. Rend. COLUMA **6,** 3 (1971).

COLE, A. W., and G. E. COATS: Tall morningglory germination response to herbicides and temperature. Weed Sci. **21,** 444 (1973).

COMSTOCK, V. E., and R. N. ANDERSEN: An inheritance study of tolerance to atrazine in a cross of flax (*Linum usitatissimum*). Crop Sci. **8,** 508 (1968).

CONNER, B. J.: Triazine herbicides and the mineral nutrition of conifers. Diss. Abstr. Int. **30,** 2483-B-84-B (1969).

——, and D. P. WHITE: Triazine herbicides and the nitrogen nutrition of conifers. Quart. Bull., Rept. Res. Michigan State Univ., Agr. Exp. Sta. **50,** 497 (1968).

COPPING, L. G.: The growth-regulator-like activity of selected s-triazines and their interactions with known plant growth substances. Diss. Abstr. Internat. **31,** 7136-B-37-B (1971).

——, and D. E. DAVIS: Effects of atrazine on chlorophyll retention in corn leaf discs. Weed Sci. **20,** 86 (1972).

—— ——, and C. G. P. PILLAI: Growth regulator-like activity of atrazine and ametryne. Weed Sci. **20,** 274 (1972).

CORBIN, F. T., J. R. BRADLEY, and H. M. MOBLEY: Pesticide interactions: Influence of phosphate insecticides on the phytotoxicity of herbicides. Pesticide interactions: Influence of p-(dimethylamino)-benzenediazo-sodium-sulfonate on the phytotoxicity of s-triazines. Proc. S. Weed Sci. Soc. **23,** 331 (1970).

COUCH, R. W.: The effect of atrazine, bromacil, and diquat on carbon^{14} fixation in corn, cotton, and soybeans. Thesis, Auburn Univ., Auburn, Ala. (1966).

——, and D. E. Davis: Effect of atrazine. bromacil and diquat on $C^{14}O_2$-fixation in corn, cotton and soybeans. Weeds **14,** 251 (1966).
Crafts, A. S.: Relation between food and herbicide transport. Isotopes Weed Res. Proc. Symp., Vienna, p. 3 (1965).
Cruz, P.: Susceptibility of five sorghum varieties to atrazine applied pre- and post-emergence. Abstr. I. Semin. de la Soc. Colombiana de Control de Malezas y Fisiologia Vegetal, p. 35 (1969).
Dalvi, R. R., B. Singh, and D. K. Salunkhe: Influence of selected pesticides on germination and associated metabolic changes in wheat and mung bean seeds. J. Agr. Food Chem. **20,** 1000 (1972).
Daris, B. T.: Selective activity against weeds of herbicides used in vineyards. Proc. Brit. Weed Control Conf., p. 785 (1968).
Davis, D. E.: Atrazine effects on mitochondrial respiration. Proc. 21st S. Weed Conf., p. 346 (1968).
——, and L. G. Copping: Growth regulator-like effects of s-triazines. Plant Physiol. **46,** Suppl. 34 (1970).
——, H. H. Funderburk, Jr., and N. G. Sansing: The absorption and translocation of C^{14}-labelled simazine by corn, cotton and cucumber. Weeds **7,** 300 (1959).
——, and C. G. P. Pillai: Effects of s-triazine herbicides on opening of hypocotyl hook of cotton. Plant Physiol. **49,** Suppl. 44 (1972).
Decleire, M., and R. Bastin: Activité de la ribulose diphosphate carboxylase extraite de froment (*Triticum aestivum*) traité avec des herbicides derivés des s-triazines. Meded. Landbouwwetenschappen **38,** 1 (1973).
——, W. De Cat and R. Bastin: Effets de traitements herbicides à base de triazine et d'urée sur la croissance, la transpiration, l'absorption des nitrates et l'activité de la nitrate réductase du froment (*Triticum aestivum*). Biochem. Physiol. Pflanzen (BPP) **165,** 175 (1974).
De Datta, S. K., W. N. Obcemea, and R. K. Jana: Protein content of rice grain as affected by nitrogen fertilizer and some triazines and substituted ureas. Agron. J. **64,** 785 (1972).
Devlin, R. M.: The influence of IAA and GA on the uptake of simazine by *Agrostis alba*. Proc. N.E. Weed Control Conf. **21,** 585 (1967).
De Vries, M. L.: The effect of simazine on monterey pine and corn as influenced by lime, bases, and aluminum sulfate. Weeds **11,** 220 (1963).
Dhillon, P. S., W. R. Byrnes, and C. Merritt: Simazine and phosphorus interaction in red pine seedlings. Weeds **15,** 339 (1967).
Diem, J. R.: Herbicidal interactions of ametryne and 2,4-D. Diss. Abstr. Internat. B **32,** 743 (1971).
——, and D. E. Davis: The enhancement of ametryne toxicity to water hyacinth with 2,4-D. Proc. S. Weed Soc., p. 359 (1969).
—— —— Enhancement of ametryne toxicity with 2,4-D. Proc. S. Weed Sci. Soc. **23,** 329 (1970).
—— —— Effect of ametryne and 2,4-D on growth and uptake of water and Ca-45 by corn and soybean. Proc. 24th Ann. Meeting S. Weed Sci. Soc., p. 351 (1971).
—— —— Effect of 2,4-D on ametryne toxicity. Weed Sci. **22,** 285 (1974).
Dill, T. R.: Effects of sub-herbicidal rates of atrazine on growth and nitrogen distribution in rye and radish seedlings. Diss. Abstr. Internat. B **34,** 2873 (1973).
——, and M. C. Carter: Distribution of nitrogen in radish and rye seedlings treated with atrazine. Abstr. Meeting Weed Sci. Soc. Amer., p. 23 (1971).
—— —— Atrazine induced changes in the distribution of N and photosynthesis. Proc. S. Weed Sci. Soc. **25,** 439 (1972).
Dimitrov, S.: Rezultati ot opiti s cherbicidi pri kartofite (Ergebnisse von Versuchen mit Herbiziden in Kartoffelbeständen). Rastenievudni nauki, Sofija **2,** 129 (1965); through Landwirt. Zentralbl. 2-66/06-1080.
Dion, A.: Chemical weed control in carrots. Rech. agron. **10,** 41 (1964/65).

DOLL, J. D.: The influence of nutrient level and combinations on herbicide uptake and phytotoxicity. Diss. Abstr. Int. **30**, 2500-B-01-B (1969).
——, and W. F. MEGGITT: Atrazine and the nitrate content in corn. Agron. J. **60**, 655 (1968).
—— —— Uptake and effectiveness of herbicides at different nutrient levels and combinations. Abstr. Meeting Weed Soc. Amer., p. 212 (1969).
——, D. PENNER, and W. F. MEGGIT: Herbicide and phosphorus influence on root absorption of amiben and atrazine. Herbicide and inorganic phosphate influence on phytase in seedlings. Weed Sci. **18**, 357 (1970).
DOMANSKA, H.: (The use of herbicides in maize crops.) Combaterea Buruienelor cu Ajutorul Erbicidelor: Referate sustinute a la Consfatuirea Internationala, Bucharest, p. 99 (1962); through Weed Abstr. **14**, 314 (1965).
DOROSHENKO, E. I., and A. F. YATSYUK: (The effect of 1,3,5-triazines on the yield and quality of fodder lupin.) Khimiya sel'. Khoz. **6**, 120 (1968); through Weed Abstr. **18**, 66 (1969).
DOROZKINA, L. A., V. A. KALININ, V. A. KUZMINSKAJA, and I. N. LEONOVA: (Der Einfluss von Atrazin and 2,4-D zusammen und einzein auf den Einbau von markiertem Kohlenstoff in Photosyntheseprodukte der Bohne.) Izvest. Timirjazevskoj sel'skochoz. Akad., Moskva **3**, 151 (1973); through Landwirt. Zentralbl. 2-74/02-1016.
DRABER, W., K. H. BUECHEL, H. TIMMLER, and A. TREBST: Structure, activity, and selectivity of 1,2,4-triazine photosynthesis inhibitors. Quantitative studies. Amer. Chem. Soc. Ser., **2**, 101 (1974).
DUKE, W. B., V. S. RAO, and J. F. HUNT: EPTC-atrazine residue interaction effect on seedling alfalfa varieties. Proc. NE Weed Sci. Soc. **26**, 258 (1972).
DUYSENS, C. N. M.: Photosynthesis. Progr. Biophys. Mol. Biol. **14**, 1 (1963).
——, and J. AMESZ: Function and identification of two photochemical systems in photosynthesis. Biochim. Biophys. Acta **64**, 243 (1962).
—— ——, and B. M. KAMP: Two photochemical systems in photosynthesis. Nature **190**, 510 (1961).
DVORAK, J.: The influence of simazine on shoot and root growth of apples and plums. Weed Res. **8**, 8 (1968).
——, and J. STANEK: Activated charcoal as a protection of strawberry plants against injury caused by simazine. Ochr. Rost. **4**, 147 (1968); through Weed Abstr. **18**, 683 (1969).
EASTIN, E. F.: Mode of action of 2-chloro-s-triazines in susceptible and resistant lines of *Zea mays* L. Thesis, Mississippi State Coll. (1963); through Weed Abstr. **13**, 823 (1964).
—— Effects of atrazine and hydroxy-atrazine on nitrogen metabolism of selected plant species. Diss. Abstr. B **27**, 2253 (1966).
—— Absorption and translocation of atrazine by resistant and susceptible lines of *Zea mays*. Abstr. Meeting Weed Sci. Soc. Amer., p. 184 (1969).
—— Growth and response to atrazine of six selections of inbred corn GT 112. Agron. J. **63**, 656 (1971).
——, and D. DAVIS: Effects of atrazine and hydroxyatrazine on nitrogen metabolism of selected species. Weeds **15**, 306 (1967).
——, R. D. PALMER, and C. O. GROGAN: Mode of action of atrazine and simazine in susceptible and resistant lines of corn. Weeds **12**, 49 (1964 a).
—— —— —— Effect of atrazine on catalase and peroxidase in resistant and susceptible lines of corn. Weeds **12**, 64 (1964 b).
EBERLE, D.: Unpublished information. Ciba-Geigy Ltd., Basle, Switzerland (1970).
EBERT, E.: Unpublished information. Ciba-Geigy Ltd., Basle, Switzerland (1968 and 1970).
——, and C. J. VAN ASSCHE: Influence of atrazine (2-chloro-4-ethylamino-6-isopropylamino-s-triazine) on auxin metabolism of plants. Experientia **25**, 758 (1969).

—— —— Effect of triazine herbicides on the physiology of plants. Summs. Paps. 7th Internat. Congress Plant Prot., Paris, p. 286 (1970).
EGAWA, H., S. TABATA, and T. NOGUCHI: Physiological effects of 2,4-dichloro-6-(orthochloroanilino)-,3,5-s-triazine (triazine, dyrene) on the chlorophylls in the leaf of rice plant. 2. Ann. Phytopath. Soc. Jap. **30**, 197 (1965).
ELENKOV, E. et al.: The effect of herbicides on the dynamics of nodule formation in French beans. Bulg. Acad. Sci. Press, Sofia, 211 (1970); through Weed Abstr. **21**, 463 (1972).
ENO, C. F.: The effect of simazine and atrazine on certain of the soil microflora and their metabolic process. Soil & Crop Sci. Soc. Florida Proc. **22**, 49 (1962).
ERCEGOVICH, C. D.: S-triazine requirement for growth of a bacterial auxotroph. Abstr. Meeting Weed Sci. Soc., p. 32 (1968).
——, and N. HERENDEEN: The utilization of s-triazines by a uracil-dependent strain of *Bacillus subtilis*. Proc. 21st NE Weed Control Conf., p. 594 (1967).
—— ——, and A. TEMPERLI: The behavior of triazine herbicides toward bacterial nucleic acids. 152nd Meet. Amer. Chem. Soc. (1966).
ESHEL, Y., and B. RUBIN: Differential tolerance of six leguminous crops to terbutryne. Israel J. Agr. Res. **22**, 11 (1972).
ETIENNE, A. L.: New results on the properties of photosystem II centers blocked by 3-(3,4-dichlorophenyl)-1,1-dimethyl-urea in their different photoactive states. Biochim. Biophys. Acta **333**, 320 (1974).
EXER, B.: Ueber Pflanzenwachstrumsregulatoren. Der Einfluss von Simazin auf den Pflanzenstoffwechsel. Experientia **14**, 135 (1958).
—— Hemmung der Hillreaktion durch Herbizide. Weed Res. **1**, 233 (1961).
FEDTKE, C.: Influence of photosynthesis-inhibiting herbicides on the regulation of crop plant metabolism. Pest. Biochem. Physiol. **2**, 312 (1972).
—— (Shade adaption reactions in wheat plants caused by the inhibitor of photosynthesis N-(2-benzothiazolyl)-N,N'-dimethylurea). Ber. deut. bot. Ges. **87**, 155 (1974).
—— Personal communication. Manuscript in preparation (1976).
FIGUEROLA, L. F.: The influence of various climatic and edaphic factors on the toxicity of 2-tert. butylamino-4-ethylamino-6-methylthio-s-triazine (IGRAN) to winter wheat (*Triticum aestivum* Vill., Host.). Diss. Abstr. B **30**, 3454 (1970).
——, and W. R. FURTICK: Edaphic factors affecting the activity of terbutryn. Weed Sci. **20**, 28 (1972 a).
—— —— Effect of climatic conditions on phytotoxicity of Terbutryn. Weed Sci. **20**, 60 (1972 b).
FINK, R. J.: The effects of atrazine or simazine-nitrogen fertilizer combinations on corn and on succeeding crops. Diss. Abstr. B **27**, 2566 (1967).
——, and O. H. FLETCHALL: The influence of atrazine or simazine on forage yield and nitrogen components of corn. Weeds **15**, 272 (1967).
FISCHER, R. A.: Stomatal opening: Rate of potassium uptake by guard cells. Science **160**, 784 (1968 a).
—— Stomatal opening in isolated epidermal strips of *Vicia faba*. II. Responses to KCl concentration and the role of potassium adsorption. Plant Physiol. **43**, 1953 (1968 b).
—— Role of potassium in stomatal opening in the leaf of *Vicia faba*. Plant Physiol. **47**, 555 (1971).
FISJUNOV, A. V.: (Der Einfluss von Herbiziden auf die Lebensfähigkeit von Unkrautsamen). Chimija v sel'skom chozjajstve, Moskva **8**, 48 (1970); through Landwirt. Zentralbl. 2-71/10-1020.
FOLAND, E. L.: Effect of some triazine herbicides on the growth and nitrogen content of young corn plants. Thesis, Univ. of Missouri (1966).
FONG, I. E., and R. F. NORRIS: Intracellular Localization of atrazine. Abstr. Meeting Weed Sci. Soc. Amer., p. 93 (1972).

FORK, D. C., and J. AMESZ: Action spectra and energy transfer in photosynthesis. Ann. Rev. Plant Physiol. **20,** 305 (1969).

FOY, C. L.: Accumulation of s-triazine derivatives in the lysigenous glands of *Gossypium hirsutum* L. Plant Physiol. **37,** 25 (1962).

——, and T. BISALPUTRA: Anatomical effects of prometryne on leaves of glanded and glandless cotton. Plant Physiol. **39,** Suppl. LXVIII (1964).

——, and H. HIRANPRADIT: Cytokinin-like activity of atrazine, bromacil, and fluometuron. Abstr. Meeting Weed Soc. Amer., p. 38 (1970).

——, and D. PENNER: Effect of inhibitors and herbicides on tricarboxylic acid cycle. Substrate oxidation by isolated cucumber mitochondria. Weeds **13,** 226 (1965).

——, S. A. TOLIN, R. A. RAGLE, and C. S. OTTO: Interactions between maize dwarf mosaic virus (MDMV) and atrazine in MDM-tolerant and MDM-susceptible varieties of corn. Abstr. Meeting Weed Soc. Amer., p. 183 (1969).

FREEMAN, J. A.: The effects of applying paraquat, atrazine and simazine for weed control in raspberries in coastal B.C. Can. J. Plant Sci. **47,** 25 (1967).

——, and A. J. RENNEY: Influence of atrazine and simazine on leaf chlorophylls and fruit yield of raspberries. Can. J. Plant Sci. **46,** 454 (1966).

FRENEY, J. R.: Increased growth and uptake of nutrients by corn plants treated with low levels of simazine. Aust. J. Agr. Res. **16,** 257 (1965).

FRIESKE, S.: (Wechsel im Gehalt von Stickstoffverbindungen und einigen mineralischen Bestandteilen in mit Simazin behandelten Pflanze). Biul. Inst. Ochrony Roslin, Poznan **48,** 19 (1971); through Landwirt. Zentralbl. 2-72/05-1089.

FUJITA, T., J. IWAZA, and C. HANCH: A new substituent constant II derived from parition coefficients. J. Amer. Chem. Soc. **86,** 5175 (1964).

FUNDERBURK, H. H., and M. C. CARTER: The effect of amitrole, atrazine, dichlobenil, and paraquat on the fixation and distribution of $C^{14}O_2$ in beans. Proc. 18th S. Weed Control Conf., p. 1 (1965).

——, and D. E. DAVIS: The metabolism of C^{14}-chain- and ring-labelled simazine by corn and the effect of atrazine on plant respiratory systems. Weeds **11,** 101 (1963).

FURTICK, W. R.: The influence of light intensity and temperature conditions on GS-14260 (2-tert. bultylamino-4-ethylamino-6-methylthio-1-triazine)-toxicity to winter wheat. Abstr. Meet. Weed Sci. Amer., p. 19 (1969).

GABBOTT, P. A.: Structure-activity relations in selected s-triazines (and acyl-anilides) inhibiting the Hill reaction. SCI-Monogr. **29,** 334 (1968).

—— Inhibition of photoreactions in isolated chloroplasts by 2-azido-4-alkylamino-6-alkylamino-s-triazines. Progr. Photosynthesis Res. **III,** 1712 (1969).

——, and G. E. BARNSLEY: 2-azido-4-(alkylamino)-6-alkylamino-s-triazines: A new group of herbicidal triazines. J. Sci. Food Agr. **19,** 16 (1968).

GAGARINA, M. U., A. S. KUPRIYANOVA, I. A. MEL'NIKOVA, and Y. A. BASKAKOV: (Herbicidal derivatives of hydroxylamine. 8. The effect of methoxy-amino-s-triazines on the Hill reaction.) Agrokhimiya **5,** 114 (1968); through Chem. Abstr. **68,** 94,792 (1968).

GARUS, I. I., I. V. KOVALEV, and R. L. LOBANOVA: (Use of triazine derivatives for suppression of weeds in agricultural fields.) Fiziol. Aktiv. Veshchestva IKh Primen. Rastenievod., Dokl. Nauch. Konf., Vilnvus, p. 73 (1963); through Chem. Abstr. **66,** 45,674 (1967).

GAST, A.: Unpublished information. Geigy Ltd., Basle, Switzerland (1956).

—— Ueber Pflanzenwachstumsregulatoren. Experientia **14,** 134 (1958).

—— Simazin als allgemeines und selektives Herbizid. 4. Internat. Pfl.'schutzkongress, Hamburg **1,** 543 (1959).

—— Neuere Erfahrungen mit Simazin im Rebbau, Schweiz. Z. Obstund Weinbau **69,** 203 (1960).

—— Beiträge zur Kenntnis des Verhaltens von Triazinen im Boden. Mededelingen von

da Landbouwhogeschool en de Opzoekings-stations van de Staat te Gent. **XXVII,** 1252 (1962).
——, and H. GROB: Triazines as selective herbicides. Pest Technol. **3,** 68 (1960).
—— —— Triazines in top fruit and viticulture. 7th Brit. Weed Control Conf., p. 217 (1964).
——, and E. KNUESLI: Ueber Pflanzenwachstumsregulatoren. Experientia **12,** 146 (1956).
GAVALAS, N. C., and H. E. CLARK: On the role of Mn^{++} in photosynthesis. Plant Physiol. **47,** 139 (1971).
GERMANO, G.: Cenni descritivi su alterazioni morfofisiologiche da diserbanti in plante ortensi. Riv. Ortoflorofrutticolture **49** (1965).
GILL, H. S.: Some effects of atrazine, amino-triazole, 2,4-D and eptam on growth, transpiration and mineral absorption in *Cyperus esculentus, Tricicum* sp. and *Zea mays.* Diss. Abstr. B **29, 1544** (1968).
——, and J. W. A. BURLEY: Effect of atrazine, amitrole, 2,4-D and eptam on the early growth of and mineral absorption by maize (*Zea mays*). Indian J. Agr. Sci. **40,** 587 (1970).
GLABISZEWSKI, J., M. PLOSZYNSKI, G. SZUMILAK, and H. ZURAWSKI: (Effect of s-triazine herbicides on oats.) Pamiet. Pulawski **21,** 233 (1966); through Chem. Abstr. **65,** 19,238 (1966).
—— —— —— —— (Influence of s-triazine herbicides on dynamics of some biochemical processes of oat in test experiments. 2. Research on the effect of propazine, pometryne and prometone.) Pamiet. Pulawski **28,** 45 (1967); through Chem. Abstr. **68,** 113,539 (1968).
GNANARETHINAM, L., and J. CARLES: La réaction physiologique de la plante à l'attaque d'un herbicide, la simazine. C. R. Soc. Biol. **160,** 1093 (1966).
GONZALEZ-IBANEZ, J.: The use of pre-emergence herbicides for weed control around young coffee seedlings. J. Agr. Univ. Puerto Rico **48,** 25 (1964).
GOOD, N. E.: Inhibitors of the Hill reaction. Plant Physiol. **36,** 78 (1961).
——, and S. IZAWA: Selective inhibitors of photosynthesis. Rec. Chem. Prog. **25,** 225 (1964).
—— ——, and G. HIND: Inhibition in photophosphorylation. Uncoupling and energy transfer. In: D. R. Sanadi (ed.): Current topics in bioenergetics. Vol. I, p. 75. New York: Academic Press (1966).
GOREN, R., and S. P. MONSELISE: Some physiological effects of triazines on citrus trees. Weeds **14,** 141 (1966).
GOSSELINK, J. G., and L. C. STANDIFER: Diurnal rhythm of sensitivity of cotton seedlings to herbicides. Science **158,** 120 (1967).
GRAESER, H.: Die Wirkung von s-Triazinen auf Wachstum und Stoffwechsel von *Zea mays.* Ber. Symposium: Wachstumsregulatoren der Pflanzen, Kühlungsborn (1966).
—— Der Einfluss von Simazin auf die löslichen Aminosäuren in etiolierten Keimlingen von *Zea mays.* Flora A **158,** 493 (1967 a).
—— Die Wirkung von s-Triazinen auf Wachstum und Stoffwechsel von *Zea mays.* Wiss. Z. Univ. Rostock Math.-Naturwiss. Reihe **16,** 565 (1967 b).
—— The influence of s-triazines on nucleic acids of *Zea mays.* 6th Meeting Fed. Eur. Biochem. Soc., Madrid, p. 900 (1969 a).
—— Influence of antagonists of protein and nucleic acid synthesis on the growth of *Zea mays* regulated by s-triazines. Biol. Zentralbl. **88,** 191 (1969 b).
—— The effect of continuous root application of simazine on the nucleic acids in etiolated seedlings of *Zea mays.* Tagungsber. Deut. Akad. Landwirtschaftswiss. Berlin. N. **109,** 97 (1970).
——, and A. GZIK: The influence of simazine on the free amino acids of etiolated maize seedlings. Tagungsber. deut. Akad. Landwirtschaftswiss. Berlin. N. **109,** 69 (1970).
——, and B. TILLICH: Die Wirkung von 5-Fluoruracil auf die Nukleinsäuren Sima-

zin-behandelter etiolierter Maiskeimlinge. Tagungsber. deut. Akad. Landwirtschaftswiss. Berlin. N. **109,** 82 (1970).

GRAHAM, J. C.: Influence of 2-chloro-4-ethylamino-6-isopropylamino-s-triazine on the transpiration of soybeans. Thesis, Univ. Wis., Madison, Wis. (1967); through Weed Abstr. **17,** 2373 (1968).

——, and K. P. BUCHHOLTZ: Alteration of transpiration and dry matter with atrazine. Weeds **16,** 389 (1968).

——, and A. ULRICH: Potassium deficiency-induced changes in stomatal behavior, left water potentials and root system permeability in *Beta vulgaris* L. Plant Physiol. **49,** 105 (1972).

GRAMLICH, J. V.: The effect of atrazine on nitrogen metabolism of resistent and susceptible species. Diss. Abstr. **26,** 649 (1965).

——, and D. E. DAVIS: Effect of atrazine on nitrogen metabolism of resistant species. Weeds **15,** 157 (1967).

—— ——, and H. H. FUNDERBURK: The effect of atrazine on nitrogen metabolism of resistant and susceptible plants. Proc. 18th S. Weed Control Conf., p. 611 (1965).

——, and R. E. FRANS: Kinetics of chlorella inhibition by herbicides. Weeds **12,** 184 (1964).

GREIG, J. K., and M. ASIF: Effects of herbicides on yield, weed control, and nutrient content of sweet corn. Proc. N. Central Weed Control Conf. **26,** 99 (1971).

GROGAN, C. O., E. F. EASTIN, and R. D. PALMER: Inheritance of susceptibility of a line of maize to simazine and atrazine. Crop Sci. **3,** 451 (1963).

GROSSBARD, E.: An appraisal of the criteria by which to measure the effect of herbicides on the soil microflora. Meded. Rijksfac. Landbouwwetensch. **35,** 515 (1970).

GROVER, R.: Germination of three coniferous species after treatment with simazine. Weeds **10,** 246 (1962).

GRUENHAGEN, R. D., and D. E. MORELAND: Effects of herbicides on ATP levels in excised soybean hypocotyls. Weed Sci. **19,** 319 (1971).

GRUZDEV, G. S., and V. A. KALININ: (Die Wirkung der Kombination von, 2,4-D mit Atrazin auf Unkräuter in Mais.) Chimija v sel'skom chozjajstve, Moskva **5,** 31 (1967); through Landwirt. Zentralbl. Abt. pflanzl. Prod. **13,** 486 (1968).

GRZESIUK, S., E. JANUSZEWICZ, and E. SOJKA: (Effect of some herbicides on the development of brewing barley, its starch content, activity of amylases and the level of gibberellins in ripening seeds.) Bull. Acad. Pol. Sci., Ser. Sci. Biol. **21,** 439 (1973 a); through Weed Abstr. **23,** 1941 (1974).

——, A. LOGIN, and J. NOWAK: (Effect of herbicides on the quality of the soluble proteins and the activity of some enzymes in germinating barley seeds.) Bull. Acad. Pol. Sci., Ser. Sci. Biol. **21,** 689 (1973 b); through Weed Abstr. **23,** 1940 (1974).

——, Z. PALIWODA, and E. SOJKA: (Effect of regular application of herbicides on the seed crop of brewing barley, its protein content and the activity of proteases.) Bull. Acad. Pol. Sci., Ser. Sci. Biol. **21,** 369 (1973 c); through Weed Abstr. **23,** 1942 (1974).

GUNTHER, F. A., and J. D. GUNTHER (ed.): The triazine herbicides. Residue Reviews **32** (1970).

——, W. C. WESTLAKE, and P. S. JAGLAN: Reported solubilities of 738 pesticide chemicals in water. Residue Reviews **20,** 1 (1968).

GYSIN, H.: Triazine herbicides—Their chemistry, biological properties and mode of action. Chem. & Ind., p. 1393 (1962).

——, and E. KNUESLI: Activity and mode of action of triazine herbicides. 4th Brit. Weed Control Conf., p. 225 (1960 a).

—— —— Chemistry and herbicidal properties of triazine derivatives. Adv. Pest Control Res. **3,** 289 (1960 b).

GZIK, A., and H. GRAESER: Der Einfluss von Simazin auf das Wachstum und den Stickstoffgehalt etiolierter Keimpflanzen von *Zea mays*. Biochem. Physiol. Pflanzen **163,** 458 (1972 a).

—— —— Untersuchungen zur Metabolisierung des Simazins in etiolierten Keimpflanzen von *Zea mays*. Biochem. Physiol. Pflanzen **163**, 483 (1972 b).

HACHIMORI, A., Y. NOSOH, and N. SONE: Effects of photosynthetic inhibitors on respiration of yeast mitochondria. J. Biochem. **64**, 119 (1968).

HAFNER, K.: Einfluss einiger wichtiger Wuchsstoffherbizide und Triazinverbindungen auf das Getreide bei vorhandener und fehlender Unkrautkonkurrenz. Thesis, Univ. Hohenheim, Germany (1963).

—— The effect of simazine on the width of the leaves of winterwheat. Z. Pflanzenkr. Pflanzenschutz, Sonderh. **II**, 1972 (1964).

HAGIMOTO, H., and H. YOSHIKAWA: Synergistic interactions between inhibitors of growth and photosynthesis. I. The "growth-dilution" hypothesis. Weed Res. **12**, 21 (1972).

HALL, A. J.: Effect of atrazine and 2,4-D applied as a preemergence treatment on *Datura ferox* and *Zea mays*. Cienc. Invest. (Buenos Aires) **23**, 79 (1967).

HAMILTON, H. R.: Tolerance of several grass species to 2-chloro-s-triazine herbicides in relation to degradation and content of benzoxazinone derivatives. J. Agr. Food Chem. **12**, 14 (1964 a).

—— A corn mutant deficient in 2,4-dihydroxy-7-methoxy-1,4-benzoxazine-3-one with an altered tolerance of atrazine. Weeds **12**, 27 (1964 b).

——, and D. E. MORELAND: Simazine degradation by corn seedlings. Science **135**, 373 (1962).

—— —— Fate of ipazine in cotton plants. Weeds **11**, 213 (1963).

HANCE, R. J.: Decomposition of herbicides in the soil by non-biological chemical processes. J. Sci. Food Agr. **18**, 544 (1967).

HANSCH, C.: Theoretical considerations of the structure-activity relationship in photosynthesis inhibitors. Progress Photosyn. Res. **III**, 1685 (1969).

——, and E. W. DEUTSCH: The structure-activity relationship in amides inhibiting photosynthesis. Biochim. Biophys. Acta **112**, 381 (1966).

——, and T. FUJITA: $\rho\sigma\pi$ analysis. A method for the correlation of biological activity and chemical structure. J. Amer. Soc. **86**, 1616 (1964).

——, A. R. STEWARD, J. IWASA, and E. W. DEUTSCH: The use of hydrophobic bonding constant for structure-activity correlations. Mol. Pharmacol. **1**, 205 (1965).

HARPER, J. L.: The evolution of weeds in relation to resistance to herbicides. Proc. 3rd Brit. Weed Control Conf., p. 179 (1956).

HARTISCH, J.: Ueber den Einfluss von Triazinen auf den Gasstoffwechsel von Kulturpflazen. Arch. Pflanzenschutz **3**, 49 (1967).

HARTMANN, W.: Wirkung von Herbiziden auf die Anthozyansynthese in Keimpflanzen von *Oenothera biennis* L. Biochem. Physiol. Pflanzen **161**, 472 (1970).

—— Ueber die Wirkungen von Herbiziden auf den pflanzlichen Stoffwechsel. Wis. Z. päd. Inst. Gustrow **9**, 19 (1971).

HASKELL, D. A., and B. J. ROGERS: The entry of herbicides into weeds. Proc. 17th NE Weed Control Conf. (1960); through Weed Abstr. **11**, 1171 (1962).

HAUCK, R. D., and H. F. STEPHENSON: Nitrification of triazine nitrogen. J. Agr. Food Chem. **12**, 147 (1964).

HAUKE-PACEWICZOWA, T.: Influence des résidus de la simazine dans le sol sur la fixation symbiotique de l'azote par les légumineuses. Meded. Rijksfac. Landbouwwetenschappen te Gent **35**, 497 (1970).

HAYES, M. H. B.: Absorption of triazine herbicides on soil organic matter, including a short review on soil organic matter chemistry. Residue Reviews **32**, 131 (1970).

HEATH, O. V. S.: Control of stomatal movement by a reduction in the normal carbon dioxide content of the air. Nature **161**, 179 (1948).

HEATHCOTE, G. D.: Weeds, herbicides and plant virus diseases. Proc. Brit. Weed Control Conf., p. 934 (1970).

HEID, R.: Eine stimulierende Wirkung durch Simazin? Allg. Forst. A. **17**, 406 (1962).

HEITEFUSS, R.: Nebenwirkungen von Herbiziden auf Pflanzenkrankheiten und deren Erreger. Z. Pflanzenkr. Pflanzenschutz Sonderh. **V**, 117 (1970).

—— Ursachen der Nebenwirkungen von Herbiciden auf Pflanzenkrankheiten. Z. Pflanzenkr. Pflanzenschutz Sonderh. **VI**, 79 (1972).

——, and H. Bodendoerfer: Der Einfluss von Herbiciden auf die Erreger von Fusskrankheiten des Getreides, insbesondere *Cercosporella herpotrichoides* und *Ophiobolus graminis*. Z. Pflanzenkr. Pflanzenschutz **77**, 109 (1970).

Helling, C. S. D., D. Kaufman, and C. T. Dieter: Algae bioassay detection of pesticide mobility in soils. Weed Sci. **19**, 685 (1971).

Hill, E. R., E. C. Putala, and J. Vengris: Atrazine-induced ultrastructural changes of barnyardgrass chloroplasts. Weeds **16**, 377 (1968).

Hill, F. R., and J. Vengris: The absorption and accumulation of atrazine in barnyardgrass in various stages of development. Proc. 22nd NE Weed Control Conf., p. 497 (1968).

Hill, K. L.: Review of structure-activity relationships between inhibitors of the Hill reaction. Proc. S. Weed Conf. **18**, 594 (1965).

Hilton, J. L., L. L. Jansen, and H. M. Hull: Mechanisms of herbicide action. Ann. Rev. Plant Physiol. **14**, 353 (1963).

Hiranpradit, H., and C. L. Foy: Retardation of leaf senescence in maize by subtoxic levels of bromacil, fluomenturon, and atrazine. Bot. Gaz. **134**, 26 (1973).

—— ——, and G. M. Shear: Effects of low levels of atrazine on some mineral constituents and forms of nitrogen in *Zea mays* L. Agron. J. **64**, 267 (1972).

Hoch, G., and R. S. Knox: Primary processes in photosynthesis. In: A. C. Giese (ed.): Photophysiology. Vol. 3, p. 225. New York: Academic Press (1968).

Hocombe, S. D.: The uptake of atrazine by germinating seeds of turnip (*Brassica rapa* L.). Weed Res. **8**, 68 (1968).

Holliday, R., and P. Putwain: Personal communication (Botany Dept., Liverpool University, 1974).

Hollister, A. T., and G. E. Walsh: Differential responses of marine phytoplankton to herbicides: Oxygen evolution. Bull. Environm. Contamin. Toxicol. **9**, 291 (1973).

Holstun, J. T., and S. W. Bingham: Several triazines as selective postemergence herbicides in cotton. Weeds **8**, 187 (1960).

Hoogstraten, S. D. van, and F. M. Ashton: Protein synthesis in vitro, as influenced by herbicides. Abstr. Meeting Weed Sci. Soc, Amer., p. 94 (1972).

Houseworth, L. D., and B. G. Tweedy: Interactions of light, temperature and moisture on terbutryn toxicity. Weed Sci. **19**, 732 (1971).

Humble, G. D., and T. C. Hsiao: Specific requirement of potassium for light-activated opening of stomata in epidermal strips. Plant Physiol. **44**, 230 (1969).

Humburg, N. E.: Modification of plant transpiration rates by herbicides. Diss. Abstr. Internat. B **31**, 5759 (1971).

Hunnius, W.: Ergebnisse der chemischen Unkrautbekämpfung bei Kartoffeln in den bayerischen Landessortenversuchen. Der Kartoffelbau **9**, 273 (1967).

Hurter, J.: Abbauprodukte von Simazine in Gramineen. Experientia **22**, 741 (1966).

—— Untersuchungen über das resistente Verhalten der Gramineen *Coix lacryma-jobi* und *Imperata cylindrica* gegenüber dem herbiziden Wirkstoff Simazin. Vjschr. naturforsch. Ges. Zürich **112**, 143 (1967 a).

—— The inactivation of simazine in resistant grasses. Abstr. 6th Internat. Congress Pl. Prot., Vienna, p. 398 (1967 b).

Ibaraki, L.: The influence of external factors on phytotoxicity of s-triazine herbicides. Weed Res., Japan **6**, 42 (1967).

Idiatulina, A. F.: (The effect of simazine on some physiological-biochemical processes in cotton.) Akad. Nauk UZ. SSR: Tashkent **4**, 22 (1963); through Chem. Abstr. **61**, 7,630 (1964).

Iliev, L.: Die Wirkung von Thioharnstoff- und s-Triazinherbiciden auf grüne und eliotierte Keimpflanzen. Thesis, Päd. Hochschule "Dr. Theodor Neubauer", Mühlhausen, Thüringen/Germany (1973).

Imbamba, S. K.: Effect of 2-chloro-4-ethylamino-6-isopropylamino-s-triazine (atra-

zine) on physiological processes in leaves, Diss. Abstr. Internat. B **31,** 1016 (1970).
——, and D. N. MOSS: Effect of atrazine on physiological processes in leaves. Crop Sci. **11,** 844 (1971).
International Rice Research Institute: Simazine and the protein content of the rice grain. Rept. Internat. Rice Res. Inst., p. 167 (1969/70).
—— Cultural practices for high protein rice. Rept. Internat. Rice Res. Inst., p. 132 (1970/71).
IVANOVA, Z. V.: (Physiologic and agrochemical principles in care of tree cultures.) Les. Khoz. **19,** 32 (1966); through Chem. Abstr. **66,** 18,228 (1967).
IWANOWA, A., and W. TSCHORBADSHIJSKA, W.: Einfluss des Herbizids Prometrin auf einige Stoffe im Gemenge von Hafer und Schnittbohne. Z. Pflanzenkr. Pflanzenschutz **74,** 378 (1967).
IZAWA, S.: High Concentration effect of Hill reaction inhibitors on isolated chloroplasts. The effect on the absorption spectra of chloroplast fragments and artificial chlorophyll a Colloid. Prog. Photosyn. Res. **III,** 1742 (1969).
——, and N. E. GOOD: The number of sites sensitive to 3-(3,4-dichloro-phenyl)-1,1-dimethylurea,3-(4-chlorophenyl-1,1-dimethylurea and 2-chloro-4-(2-propylamino-6-ethylamino-s-triazine in isolated chloroplasts. Biochim. Biophys. Acta **102,** 20 (1965).
JANCARIK, V.: Synergetic effectiveness of organic fungicides zineb and TMTD in combination with simazine against the scots pine needle-cast. Comm. Inst. forestalis Cechosloveniae, Praha, p. 229 (1965).
JANISEVSKIJ, F. V., and J. G. GRICEVIC: (Ueber Möglichkeiten der Beimischung von Propazin zu den Nitrophoska-Bestandteilen). Agrochimija **9,** 96 (1967); through Landwit. Zentralbl. 2-68/04-1084.
JANYSKA, A.: (Toleranz einiger Wurzelgemüse gegen Gesagard und Afalon). Vyzkumy Ustav Zelinarsky, Olomouc/CSSR, Bull. **9,** 55 (1965); through Landwirt. Zentralbl. Abt. pflanzl. Prod. **12,** 1364 (1967).
JESCHKE, H.-J.: Untersuchungen über den Einfluss herbizider Triazine auf Ertrag und Qualität der Kartoffel, zugleich ein Beitrag zur Methodik der Prüfung von Herbiziden in Kartoffeln. Diss. Deut. Akad. Landwirtsch.-Wiss., Berlin (1966).
JONES, L. R., B. TRUELOVE, and D. E. DAVIS: Effect of prometryne on the uptake and incorporation into protein of leucine by cucumber cotyledons. Proc. S. Weed Sci. Soc. **25,** 440 (1972).
JORDAN, L. S., and B. E. DAY: Simazine, gibberellin and light interactions on seed germination. Abstr. Meeting Weed Soc. Amer., p. 5/II (1968).
——, T. MURASHIGE, J. D. MANN, and B. E. DAY: Effect of photosynthesis-inhibiting herbicides on non-photosynthetic tobacco callus tissue. Weeds **14,** 134 (1966).
KAISER, P., J. J. POCHON, and R. CASSINI: Influence of triazine herbicides on soil microorganisms. Residue Reviews **32,** 211 (1970).
——, and H. REBER: Interactions entre la Simazine-Herbicide et les Microorganismes de la Rhizosphere du Mais. Meded. Rijksfac. Landbouwwetensch. **35,** 689 (1970).
KALIMULLINA, CH. K., and V. B. BAGAEV: (Einfluss von Propazine auf den Gehalt an Phosphorverbindungen in widerstandsfähigen und empfindlichen Pflanzen). Doklady TSChA. Moskovskaja ordena Lenina sel'skochozjajstvennaja akademija K.A. Timirjazeva **124,** 157 (1967); through Chem. Abstr. **68,** 94,803 (1968).
KALININ, F. L., and G. S. PONOMAREV: (Influence of simazine on the pigment and the carbohydrate exchange of plants.) Ukr. Botan. Zh **20,** 52 (1963); through Chem. Abstr. **61,** 15,280 (1964).
KAPPELMAN, A. J., and G. A. BUCHANAN: Influence of fungicides, herbicides and combinations on emergence and seedling growth of cotton. Agron. J. **60,** 660 (1968).
KARIM, A., and A. D. BRADSHAW: Genetic variation in simazine resistance in wheat, rape and mustard. Weed Res. **8,** 283 (1968).

KARNATZ, H.: Zur Frage der Stimulation durch Triazine bei Obstgewächsen. Z. Pflanzenkr. Pflanzenschutz Sonderh. **2**, 175 (1964).
—— Ueber die Wirkung einiger Herbizide auf Wuchs- und Ertragsleistung von Apfelbäumen. Z. Pflanzenkr. Pflanzenschutz, Sonderheft **III**, 149 (1965).
—— Untersuchungsergebnisse über den Simazingehalt in Früchten nach mehrmaliger Anwendung. Mitt. Obstbauversuchsring Alt. Land., Jorg **21**, 550 (1966).
KASASIAN, L., and C. PARKER: The effect of numerous herbicides on the germination of *Orobanche aegyptiaca* and *Striga hermontheca*. PANS **17**, 471 (1971).
KASZUBIAK, H.: (Effect of herbicides on *Rhizobium* I. Susceptibility of *Rhizobium* to herbicides.) Acta microbiol. pol. **15**, 357 (1966); through Weed Abstr. **17**, 1402 (1968).
KAUTSKY, H., W. APPEL, and H. AMANN: Chlorophyllfluorescenz und Kohlensäureassimilation. XIII. Mitteilung: Die Flouroescenzkurve und die Photochemie der Pflanze. Biochem. Z. **332**, 277 (1960).
KAY, B. L.: Atrazine and simazine increase yield and quality of range forage. Weed Sci. **19**, 370 (1971).
KAZAKOVA, K.: (Use of prometryne (A-1114) in fennel (*Foeniculum vulgare*) and anise (*Pimpinella anisum*) cultivation.) Rast. Zasht. **16**, 10 (1968); through Chem. Abstr. **72**, 53,967 (1970).
KEARNEY, P. C., and D. D. KAUFMAN (eds.): Herbicides. Vol. 1. New York and Basel: Marcel Dekker (1975).
KIERMAYER, O.: Growth responses to herbicides. In: L. Audus (ed.): The physiology and biochemistry of herbicides, p. 207. London and New York: Academic Press (1964).
KIRKLAND, K.: Reduction in atrazine toxicity to winter wheat by the growth retardant CCC. Diss. Abstr. Internat. **33**, 4066 (1973).
KLJAJIC, R., M. PLESMICAR, and V. VOJINOVIC: (A contribution to the study of harmful simazine residues in wheat plants.) Agron. Glasnik **12**, 352 (1962); through Weed Abstr. **13**, 153 (1964).
KOCH, W.: Studies on the effects of chemical substances in dormant weed seeds. Abstr. 6th Internat. Congress Plant Prot., Vienna, p. 461 (1967).
—— Untersuchungen zur Einwirkung chemischer Mittel auf dormante Unkrautsamen. Pflanzenschutzber. **38**, 167 (1968).
KOK, B., and G. M. CHENIAE: In R. Sanadi (ed.): Current topics in bioenergetics. Vol. 1, p. 1. New York: Academic Press (1966).
KONOVALOVA, V. Y.: (Application of low rates of simazine and atrazine to winter wheat.) Khimiya sel'. Khoz **9**, 41 (1967); through Chem. Abstr. **68**, 38,440 (1968).
KOZLOWSKI, T. T., and J. J. CLAUSEN: Anatomical responses of pine needles to herbicides. Nature **209**, 486 (1966).
——, and J. E. KUNTZ: Effect of simazine, atrazine, propazine and eptam on growth of pine seedlings. Soil Sci. **95**, 164 (1963).
——, and S. SASAKI: Influence of active and inert ingredients of herbicide formulations on respiration of young pine seedlings. Plant Physiol. **41**, Suppl. VII (1966).
—— —— Effects of direct contact of pine seeds or young seedlings with commercial formulations, active ingredients, or inert ingredients of triazine herbicides. Can. J. Plant Sci. **48**, 1 (1968).
—— ——, and J. H. TORRIE: Influence of temperature on phytotoxicity of triazine herbicides to pine seedlings. Amer. J. Bot. **54**, 790 (1967).
——, and J. H. TORRIE: Effect of soil incorporation of herbicides on seed germination and growth of pine seedlings. Soil Sci. **100**, 139 (1965).
KRASCHENNIKOW, N. N., and G. W. NIKITINA: (Anwendung von Prometrin in Erbsensaaten.) Chimija w sel'sk. choz. No. **2**, 32, 33 (1968); through Z. Pflanzenkr. Pflanzenschutz **76**, 233 (1969).

KRATKY, B. A., and G. F. WARREN: Reversal of the action of a photosynthetic inhibitor by activated carbon. Plant Physiol. **44**, 12 (1969).

—— —— The use of three simple, rapid bioassay on forty-two herbicides. Weed Res. **11**, 257 (1971).

KRIHNING, J.: Untersuchungen uber die Einwirkung herbizider Substanzen auf physiologische Vorgänge in Kulturpflanzen. I. Die Plasmoylse der Protoplasten der Blattstielepidermis von *Solanum tuberrosum* L. II. Der Wasserhaushalt abgeschnittner Sprosse von *Lycopersicum esculentum* Mill. III. Die Atmung und die CO_2-assimilation des Blattgewebes von *Lycopersicum escultentum* Mill. Phytopathol. Z. **53**, 65, 241, and 372 (1965).

KRINSKY, N. J.: The role of carotenoid pigments as protective agents against photosensitized oxidations in chloroplasts. In: T. W. Goodwin (ed.): Biochemistry of chloroplasts. Vol. 1, p. 423. New York: Academic Press (1967).

KRUGLOV, YU. V.: (Algalogical method of atrazine determination in soil.) Izv. Akad. Nauk. SSSR, Ser. Biol. 144 (1970); through Chem. Abstr. **73**, 2,847 (1970).

KRUGMAN, S. L., and W. B. CRITCHFIELD: Red pine needle structure. Nature **217**, 685 (1968).

KUZNETSOV, M. D.: (The effect of simazine in a fruit tree nursery.) Irv. timiryazev. sel'.-khoz. Akad. **3**, 128 (1966); through Chem. Abstr. **66**, 18,216 (1967).

KWITKO, G. P.: (Die Wirksamkeit der Herbizide in Sojasaaten.) Chimija w sel'sk. chozl **2**, 42 (1968); through Z. Pflanzenkr. Pflanzenschutz **76**, 235 (1969).

—— Effect of s-triazines on the nitrogen metabolism of wheat. Khim. Sel. Khoz. **7**, 441 (1969); through Chem. Abstr. **71**, 111,733 (1969).

LAMOUREUX, G. L., R. H. SHIMABUKURO, H. R. SWANSON, and D. S. FREAR: The identification and enzymatic synthesis of a peptide-conjugated metabolite of atrazine occurring in sorghum and corn leaves. Abstr. Papers, Amer. Chem. Soc. **157**, AGFD 16 (1969).

LARSEN, R. P., and S. K. RIES: Simazine for controlling weeds in fruit tree and grape plantings. Weeds **8**, 671 (1960).

LAVOREL, J.: Induction of fluorescence on quinone poisened chlorella cells. Plant Physiol. **34**, 2021 (1959).

—— Hétérogénéité de la chlorophylle in vivo. Biochim. Biophys. Acta **60**, 510 (1962).

LAY, M. M., and R. D. ILNICKI: Effect of atrazine on growth and nitrogen content of juvenile corn grown under two nitrate-nitrogen regimes. Proc. NE Weed Sci. Soc. **26**, 80 (1972).

LEHMANN, K., W. TUROWSKI, and S. FRIESKE: (Changes in the nitrogenous and mineral contents of plants treated with Gesatop 50.) Biuletyn Instytutu Ochrony Roslin Nr. 50, 149 (1971); through Weed Abstr. **22**, 192 (1973).

LEIDERMANN, L., and M. KRAMER: Acao de herbicidas sobre sementes de *Bideus pilosus* L. (The effect of herbicides on seeds of *Bideus pilosus* L.). ARQ Inst. Biol. (Sao Paulo) **30**, 147 (1963).

LEONARD, O. A.: Studies of monuron, diuron, simazine, and atrazine of weed control grape, quality and injury to vines. Amer. J. Enol. Vit. **12**, 69 (1961).

LEOPPKY, C., and G. B. TWEEDY: Effect of selected herbicides upon growth of soil algae. Weed Sci. **17**, 110 (1969).

LEUTHOLD, U.: Internal information, Ciba-Geigy Ltd., Basle, Switzerland (1971).

LIANG, G. H. L., K. C. FELTNER, Y. T. S. LIANG, and J. L. MORRILL: Cytogenetic effects and response of agronomic characters in grain sorghum (*Sorghum vulgare* Pens.) following atrazine application. Crop Sci. **7**, 245 (1967).

—— ——, and O. G. RUSS: Meiotic and morphological response of grain sorghum to atrazine, 2,4-D, oil and their combinations. Weed Sci **17**, 8 (1969).

——, and Y. T. S. LIANG: Effects of atrazine on chromosomal behavior in sorghum. Can. J. Genet. Cytol. **14**, 423 (1972).

LIDER, L. A., A. H. LANGE, and O. A. LEONARD: Susceptibility of grape, *Vitis vinifera*

L., varieties to root application of simazine and diuron. Proc. Amer. Soc. Hort. Sci. **88**, 341 (1966).

LIEBENOV, K.: Herbizideinfluss auf Keimung und Auflauf der wichtigsten Ackerunkräuter. Nachrbl. Deut. Pflanzenschutzdienstes **25**, 186 (1971).

LIEPINA, I.: (The effect of simazine and 2,4-D amine on the carbohydrate and nitrogen content of plants.) Latv. PSR Zinat. Akad. Vest. **5**, 73 (1969); through Chem. Abstr. **71**, 37,688 (1969).

LIN, MING-TE: The influence of simazine on nitrogen fractions of apple seedlings grown at different nitrogen levels. Thesis, Utah State Univ., Logan (1967).

——, and J. L. ANDERSON: Influence of simazine on nitrogen fractions of apple seedlings grown at different nitrogen levels. Proc. W. Weed Control Conf. **21**, 45 (1967).

LINSCOTT, D. L., and R. D. HAGIN: Protecting alfalfa seedlings from a triazine with activated charcoal. Weeds **15**, 304 (1967).

LONG, C. E., and R. F. SCRANTON: The action of charcoal on the herbicidal activity of several herbicides. Proc. N. Central Weed Control Conf. **24**, 55 (1969).

LORD, W. J., R. A. D DAMON, JR., and D. ROBINSON: Comparative responses of three apple root stocks to soil-incorporated simazine. J. Amer. Soc. Hort. Sci. **95**, 737 (1970).

LORENZONI, G. G.: Effecto stimolante della simazina a forte ciluizione (Stimulant effects of highly diluted simazine.) Maydica **7**, 115 (1962).

LOTTIKAR, P. D., L. F. REMMERT, and V. H. FREED: Effects of 2,4-D and other herbicides on oxidative phosphorylation in mitochondria from cabbage. Weed Sci. **16**, 161 (1968).

LUND-HOEIE, K.: Uptake, translocation and metabolism of simazine in Norway spruce (*Picea abies*). Weed Res. **9**, 142 (1969 a).

—— The effect of simazine on the photofixation of CO_2 and on translocation of assimilates in Norway spruce (*Picea abies*). Weed Res. **9**, 185 (1969 b).

—— Personal communication; Norwegian Forestry Res. Inst., Vollebekk (1972).

LYNCH, M. R., R. D. SWEET, and C. T. DICKERSON: Synergistic responses to atrazine in combination with other herbicides. Proc. NE Weed Control Conf. **24**, 33 (1970).

LYND, J. Q., CH. RIECK, D. BARNES, D. MURRAY, and P. W. SANTELMANN: Indicator plant aberrations at threshold soil herbicide levels. Agron. J. **59**, 194 (1967).

MAES, E., and J. STRYCKERS: The possible influence of herbicides on the baking quality of wheat. Getreide Mehl **16**, 81 (1966).

MAIER-BODE, H., and W. D. BECHTEL: Untersuchungen über Amitrol-Rückstände im Obst- und Weinbau. Z. Pflanzenkr. Pflanzenschutz, Sonderheft **IV**, 193 (1968).

MANN, D. J., H. HAID, L. S. JORDAN, and B. E. DAY: Inhibition of the action of phytochrome by the herbicide CIPC. Nature **213**, 420 (1967).

——, L. S. JORDAN, and B. E. DAY: A survey of herbicides for their effect upon protein synthesis. Plant Physiol. **40**, 840 (1965).

——, and M. PU: Inhibition of lipid synthesis by certain herbicides. Weed Sci. **16**, 197 (1968).

MANNIKAR, N. D., A. S. GILL, and C. T. ABICHANDANI: A note on effect of simazine on fodder production of m.p. chari (*Sorghum bicolor*). Cur. Sci. **40**, 641 (1971).

MARGULIES, M. M.: Effect of chloramphenicol on formation of chloroplast structure and protein during greening of etiolated leaves of *Phaseolus vulgaris*. Plant Physiol. **41**, 992 (1966).

MASHTAKOV, S. M., and R. A. PROHORCIK: (Triazine derivatives as plant growth regulators. 3. Changes in soluble carbohydrates in plants treated with simazine and atrazine.) Dokl. Akad. Nauk Beloruss. SSR **6**, 667 (1962 a); through Chem. Abstr. **58**, 8,364 (1963).

—— —— (An investigation of triazine products as regulators of plant growth. 2. The effect of simazine and atrazine on the content of chlorophyll in the leaves of crop plants.) Dokl. Akad. Nauk. Beloruss. SSR (Proc. Sci. Belorus.) **6**, 517 (1962 b); through Chem. Abstr. **58**, 3,834 (1963).

—— —— (Influence of simazine and atrazine on the growth of cultivated plants.) Fiziol. i Biokhim. Rast. Inst. Biol. Akad. Nauk Beloruss. SSR, p. 46 (1962 c); through Chem. Abstr. **59,** 5,706 (1963).

—— —— (Stimulation of photosynthesis and the Hill reaction in leaves of corn grown from seeds processed with simazine and atrazine.) Dokl. Akad. Nauk Beloruss. SSR **7,** 700 (1963 b); through Chem. Abstr. **60,** 8,569 (1964).

—— —— (Physiological reactions of plants to treatment with simazine and atrazine.) Vestsi Akad. Navuk Belarust SSR, Ser. Biyal. Navuk (3), 46 (1964); through Chem. Abstr. **62,** 11,086 (1965).

—— —— (Nature of selective effect of simazine and atrazine.) Biokhim. Osn. Zashch. Rast., Akad. Nauk SSR, Inst. Biokhim., 158 (1966); through Chem. Abstr. **66,** 7,940 (1967).

——, V. A. VOINILO, and V. P. DEEVA: (Variation in oxidative phosphorylation and the structure of the respiratory chain of *Pea mitochondria* due to action of various physiologically active compounds.) Fiziol. Rast. **16,** 111 (1969); through Chem. Abstr. **70,** 86,482 (1969).

MATTSON, R. T., R. A. KAHRS, and R. T. MURPHY: Quantitative determination of triazine herbicides in soils by chemical analysis. Residue Reviews **32,** 371 (1970).

MAZUR, T., and T. KAWECKA: (Effect of simazine on the yield and protein content of some crops.) Zesc. Nauk. Wycz. Szk. Roln. Olsztynie **25,** 925 (1969); through Chem. Abstr. **73,** 65,275 (1970).

MCDANIEL, J. L., and R. E. FRANS: Soybean mitochondrial response to prometryne and fluometuron. Weed Sci. **17,** 192 (1969).

MCGUIRE, J. J., and J. L. PEARSON: Rooting of softwood cuttings taken from container grown plants treated with simazine and diphenamid. Proc. NE Weed Sci. **26,** 62 (1972).

MCNEAL, F. H., J. M. HODGSON, and M. A. BERG: Effects of dilute sprays of simazine on spring wheat. Can. J. Plant Sci. **49,** 155 (1969).

MCREYNOLDS, W. D., JR., and J. A. TWEEDY: Effect of nitrogen form on the uptake and herbicidal activity of simazine in corn (*Zea mays* L.), rye (*Secale cereale* L.), and soybeans (*Glycine max.* L.). Abstr. Meeting Weed Soc. Amer., p. 213 (1969).

—— —— Effect of nitrogen form on simazine accumulation in corn, soybeans, and rye. Weed Sci. **18,** 270 (1970).

MEIDNER, H., and T. A. MANSFIELD: Stomatal responses to illumination. Biol. Rev. **40,** 493 (1965).

MEL'NITSCHUK, A. S., W. I. MARTYNENKO, and L. T. USCHAKOWA: (Die Wirksamkeit der Herbizide Hirsesaaten.) Chimija w sel'skom choz. **9,** 51 (1965); through Z. Pflanzenkr. Pflanzenschutz **74,** 495 (1967).

MILLER, F. R., and R. W. BOVEY: Tolerance of *Sorghum bicolor* (L) Moench. to several herbicides. Agron. J. **61,** 282 (1969).

MILLER, M. D., and D. S. MIKKELSEN: 1969 rice seed protein studies. Rice J. **73,** 38 (1970).

MILLIKAN, D . F., and DON R. MANN: Influence of the maize dwarf virus and simazine upon the buffer soluble proteins and catalase activity of a virus-susceptible and a virus-tolerant variety of *Zea mays.* Physiol. Plantarum **22,** 1139 (1969).

——, J. A. ROSS, and D. D. HEMPHILL: Influence of simazine upon the major and trace element content of soybean tissue. Abstr. Meeting Weed Soc. Amer., p. 40 (1966).

MOHS, H. J.: Untersuchungen zur Frage der Anwendung von Herbiziden im Obstbau. Thesis, Univ. Hamburg/Germany (1961).

Moldav Seed Select Inst.: SU 398-240. Priority date: 31.1.1972.

MONACO, T. J.: The partitioning and distribution of simazine and propanil in spinach chloroplasts. Thesis, N. Carolina State Univ., Raleigh (1968).

——, and D. E. MORELAND: Partitioning and distribution of herbicides in spinach chloroplasts. Abstr. Meeting Weed Soc. Amer., p. 33 (1966).

MONSON, W. G., G. W. BURTON, W. S. WILKINSON, and S. W. DUMFORD: Effect of N fertilization and simazine on yield, protein, amino-acid content, and carotenoid pigments of coastal bermudagrass. Agron. J. **63,** 928 (1971).

MONTGOMERY, M., and V. H. FREED: The uptake, translocation and metabolism of simazine and atrazine by corn plants. Weeds **9,** 231 (1961).

MOORE, J. F., and J. M. THURSTON: Inter-relationships of fungi, weeds, crops and herbicides. Proc. 10th British Weed Control Conf., p. 920 (1970).

MORELAND, D. E.: Discussion of the photochemical reactions and their properties. Proc. S. Weed Conf. **18,** 593 (1965).

—— Mechanisms of action of herbicides. Ann. Rev. Plant Physiol. **18,** 365 (1967).

—— Inhibitors of chloroplast electron transport: structure-activity relations. Progr. Photosyn. Res., **III,** 1693 (1969).

——, and K. L. HILL: The action of alkyl n-phenylcarbamates on the photolytic activity of isolated chloroplasts. J. Agr. Food Chem. **7,** 832 (1959).

—— —— Interference of herbicides with the Hill reaction of isolated chloroplasts. Weeds **10,** 229 (1962).

—— —— Inhibition of photochemical activity of isolated chloroplasts by acylanilides. Weeds **11,** 55 (1963).

——, W. A. GENTNER, J. L. HILTON, and K. L. HILL: Studies on the mechanism of herbicidal action of 2-chloro-4,6-bis(ethyl-amino)-s-triazine. Plant Physiol. **34,** 432 (1959).

——, S. S. MALHOTRA, R. D. GRUENHAGEN, and E. H. SHOKRAH: Effects of herbicides on RNA and protein synthesis. Weed Sci. **17,** 556 (1969).

MOSER, P.: Internal information. Ciba-Geigy Ltd., Basle, Switzerland (1972).

MOSS, D. N.: Photorespiration and glycolate metabolism in tobacco leaves. Crop Sci. **8,** 71 (1968 a).

—— Relation in grasses of high photosynthetic capacity and tolerance to atrazine. Crop Sci. **8,** 774 (1968 b).

MUECKE, W., P. W. MUELLER, and H. O. ESSER: Triazine nucleic acid interrelationship in *Escherichia coli* 15 arg-t-u-. Experientia **25,** 353 (1969).

MUELLER, A.: Internal information. Ciba-Geigy Ltd., Basle, Switzerland (1970).

——, E. EBERT, and A. GAST: Cytogenetic studies with atrazine (2-chloro-4-ethylamino-6-isopropylamino-s-triazine) on plants. Experientia **28,** 704 (1972).

MUELLER, J.: Vergleichende Untersuchungen der Wirkung von einigen neuen Triazinherbiziden gegenüber Koniferen und Laubhölzern in Forstbaumschulen. Arch. Forstwesen **15,** 85 (1966).

MURATA, N., M. NISHIMURA, and A. TAKAMIYA: Fluorescence of chlorophyll in photosynthetic systems. II. Induction of fluorescence in isolated spinach chloroplasts. Biochim. Biophys. Acta **120,** 23 (1966 a).

—— —— —— Fluorescence of chlorophyll in photosynthetic systems. Analysis of weak light effect in isolated chloroplasts. Biochim. Biophys. Acta **112,** 113 (1966 b).

NASYROVA, T., K. MIRKASIMOVA, and S. PAZILOVA: (Effect of some herbicides on the respiration intensity and nitrogen metabolism of cotton plants and weeds.) Usb. Biol. Zh. **12,** No. 5 (1968); through Chem. Abstr. **70,** 36,596 (1969).

NEGI, N. S., H. H. FUNDERBURK, JR., and D. E. DAVIS: Metabolism of atrazine by susceptible and resistant plants. Weeds **12,** 53 (1964).

NEWTON, M.: Christmas trees, new chemicals and techniques. Wash. Weed Control Conf., p. 47 (1966).

——, and E. S. WOODARD: Christmas trees and weed control. Oregon Weed Conf. **14,** 3 (1965).

Nippon Kayaku K.K.: JA-7225109. Priority date: 10.3.1969.

NISHIMURA, M., and A. TAKAMIYA: Energy- and electron transfer systems in algal Photosynthesis. Biochim. Biophys. Acta **120,** 45 (1966).

OKULOV, A. I., and L. G. TARNAKINA: (Die Wirkung von Simazin auf Unkräuter und Schwarze Johannisberren.) Ves. sel' skochoz. nauki, Moskva **16**, 61 (1971); through Chem. Abstr. **74**, 110,644 (1971).

OLECH, K.: Effect of some herbicides on photosynthesis and respiration of crop plants and weeds. II. Correlation between the effect of atrazine (2-chloro-4-ethylamino-6-isopropylamino-s-triazine) and external conditions. Ann. Univ. Mariae Curie-Sklodowska, Sec. E, **22**, 149 (1967); through Chem. Abstr. **71**, 11,966 (1969).

—— Influence of some herbicides on photosynthesis and respiration of crop plants and weeds. 1. Atrazine. Weed Abstr. **17**, 385 (1968).

OLIVER, L. R., and M. M. SCHREIBER: Differential selectivity of herbicides on six *Setaria* taxa. Weed Sci. **19**, 428 (1971).

OSGOOD, R. V., R. R. ROMANOWSKI, and H. W. HILTON: Differential tolerance of Hawaiian sugarcane varieties to diuron and ametryn. Abstr. Meet. Weed Soc. Amer., p. 31 (1969).

OWENS, J. R., L. G. COPPING, D. E. DAVIS, O. C. THOMPSON, and B. TRUELOVE: Plant protein synthesis and s-triazine. Effects of the s-triazine herbicides on isolated mitochondria. Effects of triazines on non-photosynthetic tissues. Proc. S. Weed Sci. Soc. **23**, 318 (1970).

PALLAS, J. E., JR., and A. R. BERTRAND: Water use, stomatal action, and transpiration unaffected by atrazine. Agron. J. **59**, 139 (1967).

PARK, R. B., and J. BIGGINS: Quantasome: Size and composition. Science **144**, 1009 (1964).

PARKS, J. R., B. TRUELOVE, and G. A. BUCHANAN: Interaction of prometryne and phorate on bean. Weed Sci. **20**, 89 (1972).

PENG, S. Y., and H. J. YEH: Determination of the varietal tolerance of sugarcane to preemergence diuron and atrazine. Weed Res. **10**, 218 (1970).

PENNER, D.: Herbicide and inorganic phosphate influence on phytase in seedlings. Weed Sci. **18**, 360 (1970).

—— Effect of disulfoton, diazinon and fensulfothion on atrazine absorption by soybean. Agron. J. **66**, 107 (1974)

——, and R. W. EARLY: Effect of atrazine on chromatin activity in corn and soybean. Weed Sci. **20**, 367 (1972).

——, and W. F. MEGGITT: Herbicide effects on soybean (*Glycine max* (L.) Merrill) seed lipids. Crop Sci. **10**, 553 (1970).

——, and L. WILEY: The influence of atrazine, benzladenine, and the embryonic axis on chlorophyll synthesis in cucumber and squash cotyledons. Physiol. Plantarum **26**, 140 (1972).

PETOJAN, E.: (Herbicide für Baumschulen). Cvedovodsto, Moskau **9**, 22 (1966); through Z 11 D-Referatekartei, Prakt. Pflanzenschutz, 66/179, Jena.

PHILLIPS, R. E., D. B. EGLI, and L. THOMPSON, JR.: Absorption of herbicides by soybean seeds and their influence on emergence and seedling growth. Weed Sci. **20**, 506 (1972).

PILET, P. E., and M. GASCHEN: A comparison of the activity of β-indoleacetic acid and of amitrole and two triazines. Rev. gen. Bot. **68**, 431 (1961); through Weed Abstr. **11**, 1160 (1962).

PILLAI, C. G. P., and D. E. DAVIS: s-Triazine effects on seed germination and hypocotyl hook opening. Weed Sci. **21**, 461 (1973 a).

—— —— Effect of s-Triazines on Lettuce Seed Germination. Abstr. Meeting Weed Sci. Soc. Amer., p. 66 (1973 b).

PINTHUS, M. J., and Y. ESHEL: Increasing the selectivity of herbicides through breeding of resistant cultivars. Photoparasitica **1**, 80 (1973).

—— ——, and Y. SHCHORI: Field and vegetable crop mutants with increased resistance to herbicides. Science **177**, 715 (1972).

PLOSZYNSKI, M.: (Investigating the inhibitory and stimulatory effect of different rates of simazine and atraton on the yields and some metabolic changes in field peas.) Pamiet. Pulaswski **51**, 51 (1972); through Weed Abstr. **22**, 1846 (1973).

—— (Influence of simazine on contents of some constituents in tops of rye plants on the background of various nitrogen fertilization.) Acta agrobot. (Warszawa) **26**, 139 (1973); through Ber. Biol. Biochem. **397**, 827 (1974).

——, and H. ZURAWSKI: (The phytotoxic action of triazine herbicides on flax, beet and buckwheat seedlings, and some physiological changes connected with it.) Acta Agrobotanica **24**, 205 (1971); through Weed Abstr. **22**, 2980 (1973).

PULVER, E. L., and S. K. RIES: Action of simazine in increasing plant protein content. Weed Sci **21**, 233 (1973).

——, and J. A. TWEEDY: Increases in nitrate and protein contents of field corn treated with simazine and atrazine. Abstr. Meeting Weed Sci. Soc. Amer., p. 30 (1968).

RADOSEVICH, S. R.: Resistance of weeds to herbicides: cause for concern. Proc. Calif. Weed Conf. **25**, 114 (1973 a).

—— Differential sensitivity of two common groundsel biotypes. Diss. Abstr. Internat. B **33**, 3428 (1973 b).

——, and A. P. APPLEBY: Differential sensitivity of two biotypes of *Senecio vulgaris* L. to several s-triazine herbicides. Res. Progr. Rept. W. Soc. Weed Sci. (1973).

—— —— Relative susceptibility to two common groundsel (Senecio vulgaris L.), biotypes to six s-triazines. Agron. J. **65**, 553 (1973 a).

—— —— Studies on the mechanism of resistance to simazine in common groundsel. Weed Sci. **21**, 497 (1973 b).

RAU, I., and L. H. GRIMME: Zum Einfluss verschieden substituierter s-Triazine auf Stoffwechselreaktionen der Grünalge *Ankistrodesmus braunii*. Z. Naturforsch. **26 B**, 919 (1971).

RENGER, G.: The watersplitting system of photosynthesis. A postulated model. Z. Naturforsch. **25 B**, 966 (1970).

—— The action of 3-(3,4-dichlorophenyl)-1,1-dimethylurea on the water splitting enzyme system Y of photosynthesis. Biochim. Biophys. Acta **314**, 113 (1973).

RHODES, M. J. C.,, and E. W. YEMM: Development of chloroplasts and the synthesis of protein in leaves. Nature **200**, 1077 (1963).

RIEDER, G., and K. P. BUCHHOLTZ: Preliminary results on the uptake of herbicides by seeds. 1968 Meeting Weed Soc. Amer., New Orleans, LA, p. 14 (1968).

—— ——, and C. A. KUST: Uptake of herbicides by soybean seed. Weed Sci. **18**, 101 (1970).

——, and E. EBERT: Der Einfluss von Triazinen auf Samen. Z. Pflanzenkr. Pflanzenschutz Sonderh. **VI**, 227 (1972).

——, and W. KOCH: Die Bestimmung der Keimpotenz von Unkrautsamen nach der TTC-Methode. Mitt. Biol. Bundesanst. f. Land- u. Forstwiss. (Berlin) **121, 210** (1967).

RIES, S. K., T. M. CHEN, and V. F. JACOUB: A comparison of toxic and sub-toxic levels of simazine on nitrogen metabolism of oats. Abstr. Meeting Weed Soc. Amer., p. 18 (1969).

——, H. CHMIEL, D. R. DILLEY, and P. FILNER: The increase in nitrate reductase activity and protein content of plants treated with simazine. Proc. Nat. Acad. Sci. USA **58**, 526 (1967 a).

——, and A. GAST: The effect of simazine on nitrogenous components of corn. Weeds **13**, 272 (1965).

——, R. P. LARSEN, and A. L. KENWORTHY: The apparent influence of simazine on nitrogen nutrition of peach and apple trees. Weeds **11**, 270 (1963).

——, O. MORENO, W. F. MEGGITT, C. J. SCHWEIZER, and S. A. ASHKAR: Wheat seed protein: Chemical influence on and relationship to subsequent growth and yield in Michigan and Mexico. Agron. J. **62**, 746 (1970).

——, C. J. SCHWEIZER, and H. CHMIEL: The increase in protein content and yield of simazine-treated crops in Michigan and Costa Rica. Biol. Sci. **18**, 205 (1968).

——, J. A. TWEEDY, H. CHMIEL, and D. R. DILLEY: Stimulation de l'activité de la réductase des nitrates et de la biosynthèse des protéines, chez les plantules de mais cultivées en solutions pauvres en nitrates et en présence de simazine. C. R. Acad. Sci. Ser. B, **53**, 868 (1967 b).

——, and V. WERT: Nitrogen balance studies with simazine treated grain species. Abstr. Meeting Weed Sci. Soc. Amer., p. **44** (1972 a).
—— —— Simazine-induced nitrate absorption related to plant protein. Weed Sci. **20**, 569 (1972 b).
ROBINSON, D. E.: Crabgrass and witchgrass susceptibility to simazine aud atrazine. Diss. Abstr. Internat. B **34**, 3590 (1974).
ROEHRBORN, G.: Chemische Konstitution und mutagene Wirkung. II. Triazinderivate. Z. Vererb'lehre **93**, 1 (1962).
ROETH, F. W.: A comparison of atrazine uptake, metabolism, and resistance in sorghum and corn. Diss. Abstr. Internat. B **31**, 1656 (1970).
ROTH, F. W.: A comparison of atrazine uptake, metabolism, and resistance in Sorghum and corn. Thesis, Univ. of Strasbourg, France (1959).
ROTH, R.: Der Einfluss einiger Herbizide auf den Geschmack der Kartoffeln. Landwirt. Versuchs- u. Untersuchungswesen **4/5**, 373 (1963); through Z. Pflanzenkr. Pflanzenschutz **74**, 493 (1967).
RUBIN, B., and Y. ESHEL: Influence of uptake and movement of terbutryn on its phytotoxicity to cotton and beans. Phytoparasitica **1**, 63 (1973).
RUBIN, S. S., and M. GRIZAJENKO: (Einfluss der Herbizide auf den anatomischen Bau der vegetativen Organe beim Mais.) Nachr. Lands.- Wiss. **11**, 42 (1964); through Pflanzenkr. Pflanzenschutz **73**, 630 (1966).
RUDGERS, L. A., J. L. DEMETERIO, G. M. PAULSON, and R. E. ELLIS, JR.: Interaction among atrazine, temperature, and phosphorus-induced zinc deficiency on corn (*Zea mays*). Soil Sci. Soc. Amer. Proc. **34**, 240 (1970).
RYAN, G. F.: Resistance of common groundsel to simazine and atrazine. Weed Sci. **18**, 614 (1970 a).
—— Resistance of common groundsel after continuous use of simazine or atrazine. Abstr. Meeting Weed Soc. Amer., p. 10 (1970 b).
SABUROVA, P. A., and A. A. PETUNOVA: [(Physiological biochemical causes of the selective action of the herbicide simazine (2-chloro-4,6-bis-ethylamino-s-triazine)]. Dokl. Akad. Nauk SSSR **160**, 1215 (1965); through Chem. Abstr. **62**, 13,778 (1965).
SALUNKHE, D. K., M. T. WU, and B. SINGH: The nutritive composition of pea and sweet corn seeds as influenced by s-triazine compounds. J. Amer. Soc. Hort. Sci. **96**, 489 (1971 a).
—— —— —— The nutritive composition of pea and sweet corn seeds as influenced by s-triazine compounds. Hort. Sci. **6**, 278 (1971 b).
SANSING, N. G., and Y. P. CHO: Lack of effect of s-triazine herbicides on nucleic acid metabolism in bacteria. Proc. S. Weed Sci. Soc. **23**, 320 (1970).
SANTELMANN, P. W., and L. EVETTS: Germination and herbicide susceptibility of six pigweed species. Weed Sci. **19**, 51 (1971).
——, C. J. SCIFRES, and J. MURRAY: Influence of postemergence herbicides on the fiber quality of selected cotton varieties. Crop Sci. **6**, 561 (1966).
SASAKI, S., and T. T. KOZLOWSKI: Influence of five herbicides on photosynthesis of pine seedlings. Plant Physiol. **40** (1965).
—— —— Influence of herbicides on respiration of young Pinus seedlings. Nature **210**, 439 (1966).
—— —— Effects of herbicides on seed germination and early seedling development of red pine. Plant Physiol. **42**, Suppl. **VII** (1967 a).
—— —— Effects of herbicides on carbon dioxide uptake by pine seedlings. Can. J. Bot. **45**, 961 (1967 b).
—— —— Effects of herbicides on seed germination and early seedling development of *Pinus resinosa*. Botan. Gaz. **129**, 2238 (1968).
—— —— Effects of ipazine (2-chloro-4-(diethylamino)-6-(isopropylamino)-5-triazine), EPTC (ethyl N,N-di-n-propylthiocarbamate), and 2,4-D on respiration of root tips of red pine (*Pinus resinosa* Ait.) plants of different ages. Adv. Front. Plant Sci. **21**, 135 (1969 a).

—— —— Effects of herbicides on respiration of red pine (*Pinus resinosa* Ait.) seedlings. 1. s-triazine and chlorophenoxy acid herbicides. Adv. Front. Plant Sci. **22**, 187 (1969 b).
—— ——, and J. H. TORRIE: Effect of pretreatment of pine seeds with herbicides on seed germination and growth of young seedlings. Can. J. Bot. **46**, 255 (1968).
SAWAMURA, S.: Cytological studies on the effect of herbicides on plant cells *in vivo*. Cytologia **30**, 325 (1965).
SAWHNEY, B. L., and I. Z. ZELITCH: Direct determination of potassium ion accumulation in guard cells in relation to stomatal opening in light. Plant. Physiol. **44**, 1350 (1969).
SCHIRMAN, R., and K. P. BUCHHOLTZ: Influence of atrazine on control and rhizome carbohydrate reserves of quackgrass. Weeds **14**, 233 (1966).
SCHROEDER, I., M. MEYER, and D. MUECKE: Die Wirkung der Herbizide 2,4-D, Amitrol, Atrazin, Chlorpropham und Chlorflurenol auf die Nucleinsäure-Biosynthese des Ascomyceten *Neurospora crassa*. Weed Res. **10**, 172 (1970).
SCHUBERT, A.: Untersuchungen über die Induktion atmungsdefekter Hefemutanten durch chemische Pflanzenschutzmittel. II. Z. allg. Mikrobiol. **9**, 483 (1969).
SCHUBERT, J., and G. WEISE: Studies on the influence of the triazine derivative simazine (W 6658) on *Sinapis alba* L. under practical conditions. Biol. Zentralbl. **87**, 629 (1968).
SCHULKE, G.: Ueber den Einfluss des Herbizides Prometryn auf das Wachstum und den Gehalt an Stickstoff von *Pisum sativum* L. Ber. Schweiz. Bot. Ges. **80**, 341 (1970).
SCHULTZ, D. P., and B. G. TWEEDY: Accumulation and incorporation of atrazine into cellular fractions of resistant and susceptible lines of *Zea mays*. Plant Physiol. **43**, Suppl. 3 (1968).
—— —— Incorporation of atrazine side-chain into amino acids of corn. Weed Sci. **19**, 133 (1971).
SCHUSTER, G.: Zur Wirkung von s-Triazinen auf Gewebe gesunder und virusinfizierter Kartoffelknollen. Biochem. Physiol. Pflanzen (BPP) **162**, 142 (1971).
SCHWEIZER, C. J., and S. K. RIES: Protein content of seed: increase improves growth and yield. Science **165**, 73 (1969 a).
—— —— Effect of protein increment in seeds by simazine and terbacil on the growth and yield of the subsequent generation. Abstr. Meeting Weed Soc. Amer., p. 182 (1969 b).
SCOTT, G. E., and C. O. GROGAN: Location of a gene in maize conditioning susceptibility to atrazine. Crop Sci. **9**, 669 (1969).
SEILER, J. P.: A survey on the mutagenicity of various pesticides. Experientia **29**, 622 (1973).
SHEETS, T. J.: Uptake and distribution of simazine by oat and cotton seedlings. Weeds **9**, 1 (1961).
SHIMABUKURO, R. H.: Atrazine metabolism in resistant corn and sorghum. Plant Physiol. **43**, 1925 (1968).
——, D. S. FREAR, H. R. SWANSON, and W. C. WALSH: Glutathione conjugation: An enzymatic basis for atrazine resistance in corn. Plant Physiol. **47**, 10 (1971).
——, and H. R. SWANSON: Atrazine metabolism in cotton as a basis for intermediate tolerance. Weed Sci. **18**, 231 (1970).
SHIMANSKII, P. S.: (The effect of 1,3,5-triazine derivatives on the mineral assimilation of pine, spruce and oak.) Agrokhimiya **7**, 120 (1970); through Chem. Abstr. **74**, 22,041 (1971).
—— (The effects of simazine and prometryne on the mineral nutrient uptake of woody plants.) Agrokhimiya **8**, 106 (1971 a); through Weed Abstr. **21**, 3184 (1973).
—— (Der Einfluss von Simazin und Prometryn auf den Verbrauch der Holzgewächse an mineralischen Nährstoffen.) Agrochimija, Moskva **12**, 106 (1971 b); through Landwirt. Zentralbl. 2-72/08-1133.

SHISHIYAMA, J., M. FUKUTOMI, and S. AKAI: Effect of some fungicides on the synthesis of chlorophylls, deoxyribonucleic acid, and ribonucleic acid in onion leaves. Phytopathol. **55**, 844 (1965).
SHONE, M. G. T., and A. V. WOOD: Factors affecting absorption and translocation of simazine by barley. J. Expt. Bot. **23**, 141 (1972 a).
—— —— Factors responsible for the tolerance of blackcurrants to simazine. Weed Res. **12**, 337 (1972 b).
SIKKA, H. C., and D. E. DAVIS: Absorption, translocation and metabolism of prometryne in cotton and soybeans. Weed Sci. **16**, 474 (1968).
—— —— Effect of prometryne on carbon-14 labeled carbon dioxide fixation in cotton and soybean. Weed Sci. **17**, 122 (1969).
—— ——, and H. H. FUNDERBURK, JR.: The effect of various types of herbicides on transpiration rate of soybean (*Glycine max.* Merr.). Proc. S. Weed Control Conf., p. 17 (1964).
SINGH, B., W. F. CAMPBELL, and D. K. SALUNKHE: Effects of s-triazines on protein and fine structure of cotyledons of bush beans. Amer. J. Bot. **59**, 568 (1972 a).
——, and D. K. SALUNKHE: Metabolic responses of bush bean plants to a subherbicidal concentration of certain s-triazine compounds. Can. J. Bot. **48**, 2213 (1970).
—— ——, and S. H. LIPTON: Growth and chemical responses of bean plants to sublethal concentrations of s-triazine compounds. J. Hort. Sci. **47**, 441 (1972 b).
——, M. T. WU, and D. K. SALUNKHE: The use of chemicals to increase leaf and seed proteins of certain crops. Plant Foods for Human Nutrition **2**, 171 (1972 c).
SINGH, B. P., G. SHANKER, and A. K. GHOSH: Preliminary study on the tolerance of germinating seeds of rough lemon and grapefruit to EPTC, Tok-E25, and simazine. Indian J. Hort. **22**, 314 (1965).
SINGH, D. J. C.: Mode of action of simazine (2-chloro-4,6-bis-(ethylamino)-s-triazine) in wheat and sunflower. Andhra Agr. J. **12**, 220 (1965).
SINGH, J. N.: Foliar absorption and translocation of prometryne in some plant species. Diss. Abstr. Internat. B **33**, 533 (1972).
——, E. BASLER, and P. W. SANTELMANN: Translocation of prometryne in several plant species. Proc. S. Weed Sci. Soc. **25**, 425 (1972).
SINGH, P. S.: The influence of simazine on metabolism of protein and ribonucleic acid in oat plants (*Avena sativa* L.). Thesis, Univ. Florida, Gainesville (1964).
SINGH, R. P., and S. H. WEST: Influence of simazine on chloroplast ribonucleic acid, and protein metabolism. Weeds **15**, 31 (1967).
SINZAR, B.: The influence of simazine on the respiration of maize and tomato plants at various growth stages. Abstr. 6th Internat. Congress Plant Prot., Vienna, p. 403 (1967).
SIVADJIAN, J., and C. P. RICARDO: Action de l'atrazine sur la transpiration foliaire. Bull. Soc. Bot. France **112**, 119 (1965).
SIVTSEV, M. V., and E. A. KUZNETSOVA: (The effect of prometryne on the condition of plastid pigments in the leaves of onion (*Allium cepa* L.). Fiziol. i Biokhimiya Kul'turnykh Rastenij **5**, 164 (1973); through Weed Abstr. **22**, 194 (1973).
SMITH, D., and K. P. BUCHHOLTZ: Transpiration rate reduction in plants with atrazine. Science **136**, 263 (1962).
—— —— Modification of plant transpiration rate with chemicals. Plant Physiol. **39**, 572 (1964).
SMITH, D. T.: Cotton yield and quality following sublethal applications of simazine and terbacil. Agron. J. **63**, 945 (1971).
SMITH, W. F., and R. D. ILNICKI: Effects of atrazine and Ca and P on fall panicum plants collected from selected sites in New Jersey. Abstr. Meeting Weed Sci. Soc. Amer., p. 24 (1972 a).
—— —— The effect of atrazine and simazine on physiological processes of fall panicum. Proc. NE Weed Sci. Soc. **26**, 79 (1972 b).

—— —— Intraspecific selectivity of atrazine to *Zoysia japonica*. 1. Toxicity studies. Proc. NE Weed Sci. Soc. **27**, 145 (1973).
——, and K. J. McVeigh: Effect of atrazine on ion uptake by corn. Proc. N.E. Weed Sci. Soc. **25**, 147 (1971).
——, and J. F. Thompson: Regulation of nitrate reductase in excised barley roots. Plant Physiol. **48**, 219 (1971).
Solecka, M., H. Profic, and D. F. Millikan: Some biochemical effects in apple leaf tissue associated with the use of simazine and amitrole. J. Amer. Soc. Hort. Sci. **94**, 55 (1969).
Solyanova, E. M.: (The effect of herbicides on cotton fibre quality and on the oil content of seeds.) Khlopkovodstvo **18**, 28 (1968); through Chem. Abstr. **69**, 18,209 (1968).
Sosnovaya, O. N.: (The effects of triazines on nutrient utilization by maize plants.) Fiz. Biokhim. kul't. Rast. **3**, 44 (1971); through Chem. Abstr. **75**, 1,470 (1971).
Spencer, D., and J. V. Possingham: The effect of manganese deficiency on photophosphorylation and oxygen-evolving sequence in spinach chloroplasts. Biochim. Biophys. Acta **52**, 372 (1961).
Spesivtsev, L. G., G. A. Chesalin, and V. F. Ladonin: (Effect of 2,4-D and atrazine on the energy metabolism of corn plants.) Khim. Sel. Khoz. **7**, 444 (1969 a); through Chem. Abstr. **71**, 111,734 (1969).
—— —— —— (Photosynthesis and respiration in relation to various concentrations of 2,4-D and atrazine.) Vest. sel'-khoz. Nauki, Mosk. **14**, 124 (1969 b); through Weed Abstr. **20**, 2754 (1971).
Standard Oil Co.: US 3551131. Priority date: 7.3.1968.
Steenbjerg, F., I. Larsen, I. Jensen, and S. Bille: The effect of nitrogen and simazine on the dry-matter yield and amino acid content of oats and on the absorption and utilization of various plant nutrients. Plant Soil **36**, 475 (1972).
Stoller, E. W., E. J. Weber, and L. M. Wax: The effects of herbicides on soybean seed constituents. J. Environ. Qua. **2**, 241 (1973).
Stolp, C. F., and D. Penner: Enhanced phytotoxicity of atrazine-phosphate combinations. Weed Sci. **21**, 37 (1973).
Stroev, V. S.: Cytogenetic activity of the herbicides simazine and maleic acid hydrazide MAH. Genetika **4**, 130 (1968 a).
—— Mutagenic effect of herbicides on barley. Genetika **4**, 164 (1968 b).
Strotmann, H., W. Tischer, and K. Edelmann: Specific binding of inhibitors by electron carriers of the photosynthetic electron transport chain. Presented German Bot. Soc., Wützburg, Sept. (1974).
Sun, C. N.: Physiological effects of the phosphorus-atrazine (2-chloro-4-ethyl-amino-6-isopropylamino-s-triazine) interaction in soybean plants (*Glycine max.*). Diss. Abstr. Internat. **31**, 994-B-95-B (1970).
——, and R. S. Adams, Jr.: Effects of the phosphorous-manganese-atrazine interaction in soybean plants. J. Agr. Food Chem. **19**, 325 (1971).
Sund, K. A., and N. Nomura: Laboratory evaluation of several herbicides. Weed Res. **3**, 35 (1963).
Susu, G. E.: (Einfluss der Herbizide auf den Eiweissgehalt im Korn und in der Grünmasse des Maises.) Kniga: Materialy IV konf. molodych ucenych Mold. Sek. fiziol., biochimii i genetiki rast., p. 167 (1966); through Chem. Abstr. **66**, 104,233 (1967).
——, L. D. Kiperman, and N. G. Nikolaeva: (Der Einfluss von Herbiziden auf die Qualität von Mais.) Tr. Kisinev, s.-ch. in-ta **76**, 96 (1970); as per Ref. Z Moskva, Rast., p. 91 (1971); through Chem. Abstr. **75**, 108,965 (1971).
Sutton, D. L., D. A. Durham, S. W. Bingham, and C. L. Foy: Influence of simazine on apparent photosynthesis of aquatic plants and herbicide residue removal from water. Weed Sci. **17**, 56 (1969).
Svrkota, B., D. Bozic, and M. Misovic: Influence of certain herbicides on the yield

and quality of potato tubers. Hrana Ishrana **9**, 10 (1968); through Chem. Abstr. **69**, 66,355 (1968).
Sweetser, P. B., C. W. Todd, and R. T. Hersh: Effect of photosynthesis inhibitors on light re-emission in photosynthesis. Biochim. Biophys. Acta **51**, 509 (1961).
Swietochowski, B., and J. Glabiszewski: (The effect of increasing rates of simazine on the rate of growth of roots and aerial parts of winter wheat.) Pamiet. Pulawski **28**, 19 (1967); through Chem. Abstr. **69**, 1,981 (1968).
——, and S. Miklaszewski: (Studies on the influence on the environment of the herbicides added to the soil. 3. Free amino acids in the soil and in oat seedlings as an index of simazine residues in the soil.) Zesz. Nauk. Wyzs. Szkoly Roln. Wroclawiu **15**, 91 (1962); through Chem. Abstr. **62**, 13,780 (1965).
——, M. Ploszynski, *and H.* Zurawski: (Effect of simazine on free amino acid and sugars in seedlings of *Avena sativa* and *Lepidium sativum*. Pamiet. Pulawski **21**, 211 (1966); through Chem. Abstr. **65**, 19,238 (1966).
—— —— —— (Investigations into the effect of different rates of simazine on some biochemical constituents of maize and quack grass, and on following crops of sorghum and oats.) Pamiet. Pulawski **31**, 93 (1968); through Chem. Abstr. **70**, 95,675 (1969).
——, M. Skowronska, and H. Zurawski: (Studies on the influence on the environment of the herbicides added to the soil. 2. The biological method of determination of simazine contents in the soil and its modification.) Zesz. Nauk. Wyzs. Szkoly Roln. Wroclawin **15**, 83 (1962); through Chem. Abstr. **62**, 13,780 (1965).
—— —— —— (Untersuchungen der Wachstumsdynamik von Hafersämlingen als Grundlage für das biologische Verfahren der Simazinbestimmung im Boden.) Zescyty naukowe wyzszey Szoly rolniczej we Wroclawiu, Rolnictwo No. 17, 305 (1964); through Landwirt. Zentralbl. Abt. pflanzl. Prod. **10**, 2660 (1965).
Tapia, M. E.: Effects of atrazine herbicide, nitrogen and phosphorous fertilizers on nutritive value of shortgrass range and blue grama grass. Diss. Abstr. Internat. B **34**, 4154 (1974).
Tarlapan, M. I., and T. V. Pancyr': (The effect of simazine on the total nitrogen, phosphorus and potassium contents in the leaves of white muscat.) Vinodelie i Vinogradarstvo **26**, 34 (1966); through Hort. Abstr. **37**, 311 (1967).
Tas, V. N.: Contribution to the study of the physiological effect of herbicides. 1. The effect of some herbicides on the germination of seed. Proc. 13th Internat. Symp. Phytofarm. Phytiatrie, Gent, p. 1465 (1961).
Temperli, A., H. Tuerler, and C. D. Ercegovich: Mode of action of pesticides. 1. Incorporation of s-triazines (cyanuric acid and prometryne) into bacterial nucleic acid. Z. Naturforsch. B **21**, 903 (1966).
Terrillon, G., and M. Paynot: Effet d'un traitement à l'atrazine sur l'activité nitrate réductase du blé. C. R. Acad. Sci., Ser. D. **277**, 2489 (1973).
Thomas, M., and H. D. Hammond: The effect of simazine and kinetin on legume modulation in *Phaseolus vulgaris* L. "red kidney." Amer. J. Bot. **55**, 721 (1968).
Thomas, V. M., L. J. Buckley, J. D. Sullivan, and Miyoshi Ikawa: Effect of herbicides on the growth of *Chlorella* and *Bacillus* using the paper disc method. Weed Sci. **21**, 449 (1973).
Thompson, L., Jr.: Metabolism of simazine and atrazine by wild cane. Weed Sci. **20**, 153 (1972 a).
—— Metabolism of chloro-s-triazine herbicides by *Panicum* and *Setaria*. Weed Sci. **20**, 584 (1972 b).
——, J. M. Houghton, F. W. Slife, and H. S. Butler: Metabolism of atrazine by Fall Panicum and Large Crabgrass. Weed Sci. **19**, 409 (1971).
——, and F. W. Slife: Environmental influence on the tolerance of corn to atrazine. Abstr. Meeting Weed Soc. Amer., p. 47 (1970).
Thompson, O. A. C.: Effects of six s-triazine herbicides on energy relations of isolated mitochondria and chloroplasts. Diss. Abstr. Internat. B **34**, 2271 (1973).

THOMPSON, O. C., and D. E. DAVIS: The effects of prometryne on energy-linked functions of mitochondria. Proc. S. Weed Sci. Soc., p. 368 (1969).
——, and B. TRUELOVE: Effects of six triazine herbicides on energy relations of subcellular organelles. Proc. S. Weed Sci. Soc. **25**, 441 (1972).
—— ——, and D. E. DAVIS: Effect of the herbicide prometryne (2,4-bis(isopropylamino)-6-(methylthio)-s-triazine) on mitochondria. J. Agr. Food Chem. **17**, 998 (1969).
—— —— —— Effects of triazines on energy relations of mitochondria and chloroplasts. Weed Sci. **22**, 164 (1974).
TIESZEN, L. L.: The effects of simazine and temperatures on photosynthesis in rye. Plant Physiol. **46**, 442 (1970).
TIMOFEVA, A. A.: (Effect of atrazine on the soluble carbohydrates in mustard plant.) Agrokhimiya 4, 123 (1967); through Chem. Abstr. **67**, 20,872 (1967).
TREBST, A.: Neue Reaktionen und Substanzen im photosynthetischen Elektronentransport. Ber. deut. bot. Ges. **83**, 373 (1970).
—— Measurement of Hill reactions and photoreduction. Methods in Enzymol. **24 B**, 146 (1972).
—— Energy conservation in photosynthetic electron transport of chloroplasts. Ann. Rev. Plant Physiol. **25**, 423 (1974).
——, and E. HARTH: Herbicidal N-alkylated-ureas and ringclosed N-acylamides as inhibitors of photosystem II. Z. Naturforsch. **29 C**, 232 (1974).
TRISKIN, S. A.: (Die Wirkung von Simazin und Monuron auf den Pflanzgutwert der Kartoffel). Chimija v sel skom chozjajstve, Moskva **6**, 39 (1968), in: Landwirt. Zentralbl. 2-69/01-0453.
TRUELOVE, B., and D. E. DAVIS: The effect of prometryne on mitochondria. Abstr. Meeting Weed Soc. Amer., p. 179 (1969).
——, L. R. JONES, and D. E. DAVIS: Light and prometryne effects on leucine uptake and incorporation. Weed Sci. **21**, 24 (1973).
TURNER, N. C.: Uptake of guard cells stimulated by fusicoccin. Nature **235**, 341 (1972).
TWEEDY, J. A., A. D. KERN, G. KAPUSTA, and D. E. MILLIS: Yield and nitrogen content of wheat and sorghum treated with different rates of nitrogen fertilizer and herbicides. Agron. J. **63**, 216 (1971).
——, and S. K. RIES: Fruit tree tolerance to two triazines. Weeds **14**, 268 (1966).
—— —— The influence of simazine on growth and nitrogen metabolism of corn. Abstr. Meeting Weed Soc. Amer., p. 60 (1967 a).
—— —— Effect of simazine on nitrate reductase activity in corn. Plant Physiol. **42**, 280 (1967 b).
UHLIG, S. K.: Untersuchungen über die Ursachen der Resistenz von Pflanzen gegenüber dem Herbizid Simazine. Thesis, Techn. Univ. Dresden (Germany) (1966 a).
—— Untersuchungen über Ursachen und Resistenz und deren Modifizierbarkeit durch Temperatur und Feuchte bei verschiedenen Baumarten gegenüber Simazin. Tharandt, Forstbot. Inst. Fak. Forstwirtsch. Plan-Nr. 35-07-52-4-14. Abschlussber. 14.4. 1966; through Landwirt. Zentralbl. pflanzl. Prod. **11**, 2243 (1966 b).
—— Studies on the problems of plant resistance to simazine. Arch. Pflanzenschutz **4**, 215 (1968).
ULRYCHOVA, M. ,and C. BLATTNY: L'action synergistique de la simazine et des viroses de plantes en tant que méthode de la détection éventuelle des maladies a virus. Biol. Plant Acad. Sci. bohemosl. **3**, 122 (1961).
UPCHURCH, R. P., G. R. LEDBETTER, and F. L. SELMAN: The interaction of phosphorus with the phytotoxicity of soil applied herbicides. Weeds **11**, 36 (1963).
VAGINA, N. S., and P. M. LATYPOVA: (Die Wirkung von Simazine auf die Gelbe Futterlupine.) Pocva, Udobrenie, Urozaj, Gorki, Sb. nauc. Tr. **88**, 155 (1972); through Landwirt. Zentralbl. 2-74/02-1024.

VAN ASSCHE, C. J.: Internal information. Ciba-Geigy Ltd., Basle, Switzerland (1971).
VAN DER ZWEEP, W.: Effects of herbicides on susceptibility of plants to pests and diseases. Proc. 10th Brit. Weed Control Conf., 917 (1970).
VAN OORSCHOT, J. L. P.: Tolerance of carrots to some herbicides inhibiting photosynthesis. XVIth Internat. Symp. Crop Prot. Ghent, 1964; Meded. Landbhogesch. Ghent **29**, 683 (1964).
—— Selectivity and physiological inactivation of some herbicides inhibiting photosynthesis. Weed Res. **5**, 84 (1965).
—— The inhibition of CO_2-uptake and transpiration of beans by diquat and semeton. Inst. Biol. Scheik. Onderz. Landbouwgewassen, Wageningen, Meded., p. 317 (1966)
—— The recovery from inhibition of photosynthesis by root-applied herbicides as an indication of herbicide inactivation. Proc. 9th Brit. Weed Control Conf., p. 624 (1968).
—— Effect of transpiration rate of bean plants on inhibition of photosynthesis by some root-applied herbicides. Weed Res. **10**, 230 (1970 a).
—— Influence of herbicides on photosynthetic activity and transpiration rate of intact plants. Pest. Sci. **1**, 33 (1970 b).
—— Photosynthetic inhibition in *Phaseolus vulgaris*. I. The analogy between herbicidal action and carbon dioxide starvation. Weed Res. **14**, 75 (1974).
——, and P. H. VAN LEEUWEN: Photosynthetic inhibition in *Phaesolus vulgaris*. II. The influence of light on herbicidal action. Weed Res. **14**, 81 (1974).
VAN OVERBEEK, J.: Physiological responses of plants to herbicides. Weeds **10**, 170 (1962).
—— Survey of mechanisms of herbicide action. In L. J. Audus (ed.): The physiology and biochemistry of herbicides, p. 387. London and New York: Academic Press (1964).
VAN RENSEN, J. J. S.: Action of some herbicides in photosynthesis of *Scenedesmus* as studied by their effects on oxygen evolution and cyclic photophosphorylation. Meded. Landbouwhogeschool Wageningen **71**, 9 (1971).
——, and P. A. VAN STEEKELENBURG: The effect of the herbicides simetone and DCMU on photosynthesis. Meded. Landbouwhogeschool Wageningen **65**, 8 (1965).
VANCE, B. D., and D. L. SMITH: Effects of five herbicides on three green algae. Texas J. Sci. **20**, 329 (1962).
VASIL'EV, D. S., and P. N. YAROSLAVSKAYA: (Prometryne control of weeds in soybeans, castor beans, mustard, and coriander plantings.) Khim. Sel. Khoz. **5**, 356 (1967); through Chem. Abstr. **68**, 38,408 (1968).
VELEV, B.: (Effect of the herbicides aretit, prometryne, and chloroxuron on the yield and some physiological processes in peas.) Rastenievud. Nauki **6**, 125 (1969); through Chem. Abstr. **72**, 110,135 (1970).
—— (Einfluss einiger Herbizide auf die Bildung von Knöllchen bei der Gemüseerbse.) Rastenievud. Nauki, Sofija 8, 129 (1971); through Landwirt. Zentralbl. 2-72/04-1062.
VERGARA, B. S., M. MILLER, and E. AVELINO: Effect of simazine on protein content of rice grain (*Oryza sativa*). Agron. J. **62**, 269 (1970).
VESELOVSKII, I. V., and YU. P. MAN'KO: (Effect of the level of mineral nutrition of plants on their resistance to simazine.) Agrokhimiya **7**, 96 (1969); through Chem. Abstr. **71**, 111,711 (1969).
VIGUE, J. T., and P. H. LI: The effect of simazine and cytokinins on potato tuber protein, nitrogen content and yield. Potato Res. **17**, 327 (1974).
VOINILO, V. A., V. P. DEEVA, and S. M. MASHTAKOV: (Changes in the oxidative phosphorylation of the mitochondria in peas owing to the action of chemically different herbicides.) Dokl. Akad. Nauk. Beloruss., SSR **11**, 638 (1967); through Chem. Abstr. **70**, 86,482 (1969).

VOLODIN, V. S.: (Economic efficiency of herbicides on grey forest soils of the central chernozem zone.) Vest. sel'.-khoz. Nauki, Mosk. **13**, 82 (1968); through Weed Abstr. **18**, 1561 (1969).
VOROB'EV, F. K., and J.-P. CH'A: (Effect of simazine and 2,4-D on nitrogen metabolism of plants.) Dokl. Moskov. Sel'skhokhoz. Akad. K. A. Timirjazeva **57**, 63 (1960); through Weed Abstr. **11**, 1579 (1962).
WAGGONER, P. E., and I. ZELITCH: Transpiration and the stomata of leaves: Water-loss through leaf pores is controlled by pore size, which varies with environment and chemical sprays. Science **150**, 1413 (1965).
WAKONIG, R., and J. T. ARNASON: Effect of triazines on chromosomes. Proc. 10th Internat. Congress Genetics, Montreal **2**, 205 (1958).
—— —— Chromosome breakage induced in *Vicia faba* root tips by 2,4,6-tri(ethyleneimino)-1,3,5-triazine. Can. J. Bot. **37**, 403 (1959).
WALKER, D. A., and I. ZELITCH: Some effects of metabolic inhibitors, temperature and anaerobic conditions on stomatal movement. Plant Physiol. **38**, 390 (1963).
WALSH, G. E.: Effects of herbicides on photosynthesis and growth of marine unicellular algae. Hyacinth Control J. **10**, 45 (1972).
WARWICK, D. D.: The feasibility of breeding for herbicide resistance in a crop plant. Thesis, Univ. of Liverpool, Eng. (1973).
WATERS, W. E.: The influence of diuron and simazine herbicides on bulb production of *Lilium longiflorum*. Florida State Hort. Soc., p. 424 (1963).
WEIDNER, C, W., and T. L. LAVY: Use of an algae bio-assay for determining trace amounts of atrazine, alachlor, butylate, and picloram in water. Res. Rept. N. Central Weed Control Conf., p. 30 (1973).
WELKER, W. V., JR., and J. L. BROGDON: Response of highbush blueberries to long-term use of diuron and simazine. Weed Sci. **16**, 303 (1968).
WESSELS, J. S. C., and R. VAN DER VEEN: The action of some derivatives of phenylurethan and of 3-phenyl-1,1-dimethylurea on the Hill reaction. Biochim. Biophys. Acta **19**, 548 (1956).
WETTASINGHE, D. T.: A preliminary investigation of the effect of two plant factors on simazine toxicity. Proc. Brit. Weed Control Conf. **9**, 645 (1968).
WHITENBERG, D. C.: The fact of prometryne in cotton plants. Weeds **13**, 68 (1965).
WIEDMAN, S. J.: Stimulation of plant growth by sub-lethal concentrations of herbicides. Diss. Abstr. Internat. **30**, 4464-B-65-B (1970).
——, and A. P. APPLEBY: Plant growth stimulation by sublethal concentration of herbicides. Weed Res. **12**, 65 (1972).
WIESE, A. F., T. J. ARMY, and J. D. THOMAS: Moisture utilization by plants after herbicide treatment. Weeds **14**, 205 (1966).
——, and J. R. QUINBY: Inheritance of 2,4-D and propazine resistance in grain sorghum. Abstr. Meeting Weed Soc. Amer. **29** (1969).
WILKINSON, R. E., and W. S. HARDCASTLE: Herbicide influence on cottonseed oil quality. J. Agr. Food Chem. **19**, 851 (1971).
—— —— Influence of herbicides as single applications or mixtures on fatty acid compositions of cottonseed oil. J. Agr. Food Chem. **20**, 996 (1972 a).
—— —— Cotton oil quality after postemergence herbicides in oil. Weed Sci. **20**, 241 (1972 b).
WILLS, G. D., and D. E. DAVIS: The influence of atrazine on the water uptake of plants. Proc. 15th S. Weed Conf., p. 270 (1962).
—— ——, and H. H. FUNDERBURK: The effect of atrazine on transpiration in corn, cotton, and soybeans. Weeds **11**, 252 (1963).
WONG, M. K., and R. R. ROMANOWSKI, JR.: A relationship between foliar symptoms and tissue content of cucumber sprayed with atrazine. Weed Sci. **16**, 441 (1968).
WORLEY, R. R., and S. A. HARMON: No simazine residue in pecan nuts. Hort. Sci. **2**, 150 (1967).
WRAY, J. L., S. K. RIES, and P. FILNER: Protein accumulation and regulation of

nitrate reduction in higher plants. In: FAO/IAEA Symposium (Vienna), Plant protein resources, their improvements through nuclear techniques **1**, 403 (1970).
WU, C. H., P. W. SANTELMANN, and J. M. DAVIDSON: Influence of soil temperature and moisture on terbutryn activity and persistence. Weed Sci. **22**, 571 (1974).
WU, M. T., D. K. SALUNKE, and B. SINGH: Metabolic responses of pea and sweet corn plants to sublethal concentrations of s-triazine compounds. Proc. Amer. Soc. Plant Physiol., p. 287 (1971 a).
——, B. SINGH, and D. K. SALUNKHE: Influence of foliar application of simazine, igran and GS-14254 on delta-aminolevulinic acid dehydratase of pea seedlings. Phytochem. **10**, 2025 (1971 b).
—— —— —— Influence of s-triazines on some enzymes of carbohydrates and nitrogen metabolism in leaves of pea (*Pisum sativum* L.) and sweet corn (*Zea mays* L.). Plant Physiol. **48**, 517 (1971 c).
—— —— —— Effects of s-triazines on protein synthesis in leaves of peas (*Pisum sativum* L.) and sweet corn (*Zea mays* L.) and on the ultrastructure of pea cotyledons. Experientia **28**, 1002 (1972 a).
—— —— —— Influence of foliar application of s-triazine compounds on fresh weight, dry weight, chemical composition, and enzymatic activity of pea and sweet corn seedlings. J. Expt. Bot. **23**, 793 (1972 b).
WUU, K. D. and W. F. GRANT: Morphological and somatic chromosomal aberrations induced by pesticides in barley (*Hordeum vulgare*). Can. J. Genet. Cytol. **8**, 481 (1966).
—— —— Chromosomal aberrations induced by pesticides in meiotic cells of barley. Cytologia **32**, 31 (1967 a).
—— —— Chromosomal aberrations induced in somatic cells of *Vicia faba* by pesticides. The Nucleus **10**, 37 (1967 b).
ZAHN, H.: Versuche zur Anwendung von Herbiziden in Hohn. Herba hung. **5**, 190 (1966).
ZAVARZIN, V. I., and T. V. BELJAEVA: (Herbicides and fruit quality.) Sadovodstvo No. 5, 13 (1968); through Hort. Abstr. **39**, 227 (1969).
ZAVITKOVSKI, J., and M. NEWTON: Effect of two herbicides on germination and survival of several coniferous and herbaceous species. W. Weed Control Conf. Res. Rept. (1964).
ZELEV, A., P. PAVLOV, A. DIMOV, and R. KONDAREV: (Nachwirkungen einiger Triazinpräparate auf den Ertrag und den Saatgutwert des Weizens.) Naueni trudove. Visselskostopanski inst. "Vasil Kolarov" Kn. 1. Agron. Fak. Plovdiv **16**, 97 (1967); through Landwirt. Zentralbl. 2-69/01-1102.
ZENK, M. H.: Haploids in physiological and biochemical research. In K. J. Kasha (ed.): Haploids in higher plants, p. 339. Univ. of Guelph, Canada (1974).
ZELITCH, I.: Stomatal control. Ann. Rev. Plant Physiol. **20**, 329 (1969).
ZHIRMUNSKAYA, N. M., and G. S. JOFFE: (Effect of atrazine on potassium absorption by barley roots.) Fiziol. Rastenij **20**, 482 (1973); through Chem. Abstr. **79**, 62, 526 (1973).
——, and S. S. KOLTZOVA: (A study on permeability of barley root cells for herbicide atrazine. Fiziol. Rastenij **20**, 151 (1973); through Ber. Biochem. Biol. **388**, 836 (1974).
ZINCHENKO, V. A., L. S. FURSENKO, and T. V. OSINSKAYA: (The effect of triazine derivatives on the mineral and water content of horse bean plants.) Agrokhimiya, p. 103 (1966); through Chem. Abstr. **65**, 17,617 (1966).
ZUMFT, W.: Personal communication: Univ. of Erlangen, Germany (1970).
ZWATZ, B., and K. WALTL: Untersuchungen über die Beeinflussung der Qualitat von Winterweizen durch Anwendung von Fungiziden und Herbiziden. Pflanzenschutz Ber. **44**, 69 (1974).
ZWEIG, G., and F. M. ASHTON: The effect of atrazine on distribution of ^{14}C-compounds following $^{14}CO_2$-fixation in excised kidney bean leaves. J. Expt. Bot. **13**, 5 (1962).

——, I. TAMAS, and E. GREENBERG: The effect of photosynthesis inhibitors on oxygen evolution and fluorescence of illuminated chlorella. Biochim. Biophys. Acta **66**, 196 (1963).

Manuscript received May 19, 1975; accepted February 16, 1976.

Appendix

Chemical identification of common names and abbreviations used in text.

Common names and abbreviations	*Chemical names*
ADP	adenosine diphosphate
ametryn	2-ethylamino-4-isopropylamino-6-methylthio-*s*-triazine
Ammeline	4,6-diamino-2-hydroxy-*s*-triazine
ATP	adenosine triphosphate
Atraton	2-ethylamino-4-isopropylamino-6-methoxy-*s*-triazine
Atrazine	2-chloro-4-ethylamino-6-isopropylamino-*s*-triazine
BOA	benzoxazinone
CCC	(2-chloroethyl)-trimethyl-ammonium chloride
Chlorazine	2-chloro-4,6-bis(diethylamino)-*s*-triazine
Cyprazine	2-chloro-4-cyclopropylamino-6-isopropylamino-*s*-triazine
DCMU	3-(3,4 dichlorophenyl)-1,1-dimethylurea
FMN	flavin mononucleotide
GA	gibberellic acid
GS-14254	2-*sec*.butylamino-4-ethylamino-6-methoxy-*s*-triazine
Hydroxyatrazine	2-ethylamino-4-hydroxy-6-isoproplyamino-*s*-triazine
Hydroxysimazine	2,4-bis(ethylamino)-6-hydroxy-*s*-triazine
IAA	3-indolyl-acetic acid
Ipazine	2-chloro-4-diethylamino-6-isopropylamino-*s*-triazine
Melamine	2,4,6-triamino-*s*-triazine
Metoprotryn	2-isopropylamino-4-(γ-methoxypropylamino)-6-methylthio-*s*-triazine
NAA	α-naphthyl acetic acid
PMS	phenazine-methosulfate
Prometon	2,4-bis(isopropylamino)-6-methoxy-*s*-triazine
Prometryn	2,4-bis(isopropylamino)-6-methylthio-*s*-triazine
Propazine	2-chloro-4,6-bis(isopropylamino)-*s*-triazine
Simazine	2-chloro-4,6-bis(ethylamino)-*s*-triazine
Simeton	2,4-bis(ethylamino)-6-methoxy-*s*-triazine
Simetryn	2,4-bis(ethylamino)-6-methylthio-*s*-triazine
Terbutryn	2-*tert*.butylamino-4-ethylamino-6-methylthio-*s*-triazine

Subject Index

GPSR Compliance

The European Union's (EU) General Product Safety Regulation (GPSR) is a set of rules that requires consumer products to be safe and our obligations to ensure this.

If you have any concerns about our products, you can contact us on ProductSafety@springernature.com

In case Publisher is established outside the EU, the EU authorized representative is:

Springer Nature Customer Service Center GmbH
Europaplatz 3
69115 Heidelberg, Germany

Batch number: 10704707

Printed by Printforce, the Netherlands